Ernst Schering Research Foundation Workshop 16
Organ-Selective Actions of Steroid Hormones

Springer-Verlag Berlin Heidelberg GmbH

Ernst Schering Research Foundation
Workshop 16

Organ-Selective Actions of Steroid Hormones

D. T. Baird, G. Schütz, R. Krattenmacher
Editors

With 44 Figures and 8 Tables

 Springer

ISBN 978-3-662-09155-5

Die Deutsche Bibliothek – CIP-Einheitsaufnahme
Ernst-Schering-Forschungsgesellschaft <Berlin>:
Ernst Schering Research Foundation workshop.

Früher Schriftenreihe. – Früher angezeigt u.d.T.: Ernst-Schering-Forschungsgesellschaft
<Berlin>: Schering Foundation workshop. – Teilw. u.d.T.: Ernst-Schering-Forschungs-
gesellschaft <Berlin>: Schering Foundation workshop
ISSN 0947-6075
NE: Ernst-Schering-Forschungsgesellschaft <Berlin>; Schering Foundation workshop;
HST
16. Organ selective actions of steroid hormones. – 1995
Organ selective actions of steroid hormones / D.T. Baird ... ed.

(Ernst Schering Research Foundation workshop; 16)
ISBN 978-3-662-09155-5   ISBN 978-3-662-09153-1 (eBook)
DOI 10.1007/978-3-662-09153-1
NE: Baird, David T. [Hrsg.]

CIP data applied for

The use of general descriptive names, registered names, trademarks, etc. in this publica-
tion does not imply, even in the absence of a specific statement, that such names are
exempt from the relevant protective laws and regulations and therefore free for general
use. Product liability: The publishers cannot guarantee the accuracy of any information
about dosage and application contained in this book. In every individual case the user
must check such information by consulting the relevant literature.

Typesetting: Data conversion by Springer-Verlag

21/3135–5 4 3 2 1 0 – Printed on acid-free paper

Preface

Pharmacodynamics can be defined as the study of the biochemical and physiological effects of drugs and mechanisms of their action. The latter aspect of the subject is perhaps the most fundamental challenge to the investigator in pharmacology, and information derived from such studies is useful for the clinician. The objectives of the analysis of drug action are to identify the primary action, to delineate the chemical or physical interactions between drug and cell and to characterize the full sequence and scope of actions and effects. Such a complete analysis provides the basis for both the rational therapeutic use of a drug and the design of new therapeutic agents.

The last decade has witnessed a vast extension and coalescence of our knowledge of the structures, mechanism of action, and biochemical functions of steroid hormone receptors. Steroid receptors and many other proteins that are involved in biological signal transduction have been purified. Molecular cloning has provided amino acid sequences for steroid hormone receptors, permitted their expression and study in genetically defined backgrounds and allowed a detailed analysis of structure–function relationships of signaling molecules through site-directed mutagenesis.

Due to the specific expression of steroid hormone receptors in different target tissues, organ-specific action is not a new feature of steroids. However, the rapidly increasing knowledge of the complex set of intracellular interactions opens numerous further options for organ specificity. For example, in epithelial cells of the distal nephron, aldosterone can occupy its receptor exclusively only when the enzyme 11βHSD, which is colocalized with the mineralocorticoid receptor, locally inactivates glucocorticoids.

Fig. 1. The participants of the workshop. *From left to right, background*:
R. L. Sutherland, R. T. Turner, D. T. Baird, G. Schütz, M. Bygdeman;
foreground: J. S. Finkelstein, D. P. McDonnell, B. S. Katzenellenbogen,
H. F. DeLuca, M. C. Farach-Carson, R. Krattenmacher

It would necessarily be beyond the scope of this workshop to sum-
marize all aspects of organ-specific actions of steroid hormones. Our
aim was rather to highlight and discuss recent findings concerning
basic molecular mechanisms of organ specificity and to bring them
into close relationship with clinical aspects. The contributions of the

leading experts in steroid physiology who participated in this work-shop from May 17–19, 1995 will lead to a better understanding of the underlying mechanisms of organ selectivity and help towards improved therapies by taking advantage of tissue specificity.

The editors gratefully acknowledge the contributions of the authors of the chapters in this book and the assistance provided by the Ernst Schering Research Foundation, in particular by Dr. U.F. Habenicht and Mrs U. Wanke.

D.T. Baird
G. Schütz
R. Krattenmacher

Table of Contents

1 Development of Tissue-Selective Estrogen Receptor
Modulators
D. P. McDonnell, B. A. Lieberman, and J. Norris 1

2 Estrogen-Receptor and Antiestrogen-Receptor Complexes:
Cell- and Promoter-Specific Effects and Interactions
with Second Messenger Signaling Pathways
*B. S. Katzenellenbogen, M. M. Montano, W. L. Kraus,
S. M. Aronica, N. Fujimoto, and P. LeGoff* 29

3 Analysis of Genetically Altered Mice
Without Glucocorticoid Receptor
W. Schmid, T. Cole, J. Blendy, L. Montoliu, and G. Schütz . 51

4 Organ-Selective Actions of Tamoxifen
and Other Partial Antiestrogens
R. T. Turner 65

5 Progestin Regulation of Cell Proliferation
in the Breast and Endometrium
*R. L. Sutherland, C. S. L. Lee, A. L. Cornish,
and E. A. Musgrove* 85

6 Central Versus Endometrial Effects of Antiprogestins:
Is Endometrial Selectivity Possible?
M. Bygdeman, K. Gemzell-Danielsson, and M. L. Swahn . . 107

7   Androgen Action on the Bone
     J. S. Finkelstein 121

8   General Principles of Vitamin D Action and Mechanism-Based
     Search for Analogs with Specific Actions
     H. F. DeLuca, C. Zierold, and H. M. Darwish 137

9   Organ-Specific Actions of Vitamin D Analogs:
     Relevance of Rapid Effects
     M. C. Farach-Carson and S. E. Guggino 161

Subject Index 181

Previous Volumes Published in this Series 183

List of Editors and Contributors

Editors

D. T. Baird
Department of Obstetrics and Gynaecology, Centre for Reproductive Biology,
University of Edinburgh, 37 Chalmers Street, Edinburgh EH3 9EW, Scotland

G. Schütz
German Cancer Research Center, Molecular Biology of the Cell I,
Im Neuenheimer Feld 280, 69120 Heidelberg, Germany

R. Krattenmacher
Research Laboratories of Schering AG, 13342 Berlin, Germany

Contributors

S. M. Aronica
Department of Physiology and Biophysics, University of Illinois
and University of Illinois College of Medicine, 524 Burrill Hall,
407 South Goodwin Avenue, Urbana, IL 61801-3704, USA

J. Blendy
German Cancer Research Center, Molecular Biology of the Cell I,
Im Neuenheimer Feld 280, 69120 Heidelberg, Germany

M. Bygdeman
Department of Woman and Child Health, Division for Obstetrics
and Gynecology, Karolinska Hospital, 17176 Stockholm, Sweden

T. Cole
German Cancer Research Center, Molecular Biology of the Cell I,
Im Neuenheimer Feld 280, 69120 Heidelberg, Germany

A. L. Cornish
Cancer Biology Division, Garvan Institute of Medical Research,
St. Vincent's Hospital, Darlinghurst, NSW, 2010, Australia

H. M. Darwish
College of Agricultural and Life Sciences, University of Wisconsin-Madison,
420 Henry Mall, Madison, WI 53706, USA

H. F. DeLuca
College of Agricultural and Life Sciences, University of Wisconsin-Madison,
420 Henry Mall, Madison, WI 53706, USA

M. C. Farach-Carson
Department of Basic Sciences, Section of Biochemistry, University of Texas-
Houston, Dental Branch, 6516 John Freeman Avenue, Houston, TX 77030,
USA

J. S. Finkelstein
Massachusetts General Hospital, Endocrine Unit, Bulfinch 327,
32 Fruit Street, Boston, MA 02114, USA

N. Fujimoto
Department of Physiology and Biophysics, University of Illinois
and University of Illinois College of Medicine, 524 Burrill Hall,
407 South Goodwin Avenue, Urbana, IL 61801-3704, USA

K. Gemzell-Danielsson
Department of Woman and Child Health, Division for Obstetrics
and Gynecology, Karolinska Hospital, 17176 Stockholm, Sweden

S. E. Guggino
Department of Medicine, Division of Gastroenterology,
Johns Hopkins School of Medicine, Baltimore, MD 21205, USA

B. S. Katzenellenbogen
Department of Physiology and Biophysics and Department of Cell
and Structural Biology, University of Illinois and University of Illinois
College of Medicine, 524 Burrill Hall, 407 South Goodwin Avenue,
Urbana, IL 61801-3704, USA

W. L. Kraus
Department of Physiology and Biophysics, University of Illinois
and University of Illinois College of Medicine, 524 Burrill Hall,
407 South Goodwin Avenue, Urbana, IL 61801-3704, USA

C. S. L. Lee
Cancer Biology Division, Garvan Institute of Medical Research,
St. Vincent's Hospital, Darlinghurst, NSW, 2010 Australia

P. LeGoff
Department of Physiology and Biophysics, University of Illinois
and University of Illinois College of Medicine, 524 Burrill Hall,
407 South Goodwin Avenue, Urbana, IL 61801-3704, USA

B. A. Lieberman
Duke University Medical School, Department of Pharmacology, Box 3813,
Durham, NC 27710, USA

D. P. McDonnell
Duke University Medical School, Department of Pharmacology, Box 3813,
Durham, NC 27710, USA

M. M. Montano
Department of Physiology and Biophysics, University of Illinois
and University of Illinois College of Medicine, 524 Burrill Hall,
407 South Goodwin Avenue, Urbana, IL 61801-3704, USA

L. Montoliu
German Cancer Research Center, Molecular Biology of the Cell I,
Im Neuenheimer Feld 280, 69120 Heidelberg, Germany

E. A. Musgrove
Cancer Biology Division, Garvan Institute of Medical Research,
St. Vincent's Hospital, Darlinghurst, NSW, 2010, Australia

J. Norris
Duke University Medical School, Department of Pharmacology, Box 3813,
Durham, NC 27710, USA

W. Schmid
German Cancer Research Center, Molecular Biology of the Cell I,
Im Neuenheimer Feld 280, 69120 Heidelberg, Germany

G. Schütz
German Cancer Research Center, Molecular Biology of the Cell I,
Im Neuenheimer Feld 280, 69120 Heidelberg, Germany

R. L. Sutherland
Cancer Biology Division, The Garvan Institute of Medical Research,
St. Vincent's Hospital, 384 Victoria Street,
Darlinghurst Sidney, NSW, 2010, Australia

M. L. Swahn
Department of Woman and Child Health, Division for Obstetrics
and Gynecology, Karolinska Hospital, 17176 Stockholm, Sweden

R. T. Turner
Department of Orthopedic Research, Mayo Clinic and Foundation,
200 First Street SW, Rochester, MN 55905, USA

C. Zierold
College of Agricultural and Life Sciences, University of Wisconsin-Madison,
420 Henry Mall, Madison, WI 53706, USA

1 Development of Tissue-Selective Estrogen Receptor Modulators

D. P. McDonnell, B. A. Lieberman, and J. Norris

1.1	Introduction	1
1.2	The Mechanism of Action of ER Resembles That of Other Steroid Hormone Receptors	2
1.3	Identification of Cell-Specific ER Modulators	4
1.4	Specific Domains Within ER Interact with the Cellular Transcription Machinery and Mediate the Transcriptional Effects of Estradiol	5
1.5	The Pharmacology of ER Ligands Is Influenced by the Transcriptional Activity of TAF-1 and TAF-2	9
1.6	The Agonist Activity of Estradiol and Tamoxifen Can Be Distinguished by Specific ER Mutations	14
1.7	Identification of Cellular Factors that Interact with and Modify the Biological Activity of the TAFs Within ER	17
1.8	A Model Describing the Pharmacology of ER Modulators	21
1.9	Final Comments	23
References		24

1.1 Introduction

The steroid hormone estrogen is a key intracellular modulator of the processes involved in differentiation, homeostasis, and development of female reproductive function (Clark and Peck 1979). In pathological states, estrogen is also involved in the maintenance and progression of cancers (Sunderland and Osborne 1991) and is implicated in the abnormalities of uterine function observed in endometriosis and possibly uterine fibroids (Kettel et al. 1991). The pharmaceutical exploitation of the estrogen receptor (ER) has seen the development of antihormones, compounds which can oppose the function of the natural hormone estrogen (Kedar et al. 1994; Sunderland and Osborne 1991). One such compound, tamoxifen, has found widespread use in the clinic as adjuvant hormonal therapy for breast cancer where it functions as an antagonist of those estrogen-regulated genes responsible for cellular proliferation (Kedar et al. 1994; Sunderland and Osborne 1991). Interestingly, in bone and in the cardiovascular system, tamoxifen exhibits sufficient ER agonist activity to maintain bone mass and cardiovascular tone in postmenopausal women (Love et al. 1992). Because of these favorable activities, along with its breast-selective antagonist activities, tamoxifen is currently being evaluated as a chemopreventative agent in patients who are at high risk for developing breast cancer (Henderson et al. 1993). Unfortunately, its utility for this application may be adversely affected according to a recent study which indicated that 37% of women taking tamoxifen demonstrated histological changes in the endometrium indicative of unopposed ER agonist activity (Kedar et al. 1994). Whether these alterations reflect a carcinogenic potential of tamoxifen is unclear. Nevertheless the finding that tamoxifen displays tissue-selective ER agonist and antagonist activities in humans strongly supports the concept that different compounds working through the same receptor can direct a distinct pattern of gene expression. Understanding the mechanism by which tamoxifen manifests tissue-restricted biological activity will facilitate the development of additional ER modulators with improved therapeutic profiles. It is also likely that these insights will extend to other steroid receptor-mediated transduction pathways, thus allowing the development of novel tissue (or pathway) specific progestins, androgens, glucocorticoids, or mineralocorticoids.

1.2 The Mechanism of Action of ER Resembles That of Other Steroid Hormone Receptors

The availability of the cloned cDNAs for the known steroid receptors and their use to reconstitute hormone responsive systems in heterologous cells has permitted a detailed dissection of the steroid hormone receptor signal transduction pathways. These studies have revealed that the mechanism by which ER and other steroid hormone receptors mediate their biological effects in target cells is similar (Beato 1989; McDonnell et al. 1993).

In the absence of hormone, the latent receptor resides in target cell nuclei as part of a large macromolecular complex comprising heat-shock protein 90 (hsp90), hsp70, p59 and other proteins (Smith and Toft 1993). No clear function has yet been identified for the heat-shock proteins in ER action. However, it has been shown in vitro that the glucocorticoid receptor (GR) requires hsp interaction to acquire its hormone-binding activity, possibly by assisting in the correct folding of the ligand-binding domain (Pratt et al. 1992). In addition, it has been suggested that hsp interaction may be required to maintain the steroid hormone receptors transcriptionally inactive in the absence of hormone (Sanchez et al. 1990; Smith and Toft 1993). Several studies have demonstrated, using sensitivity to protease digestion as an assay, that ligand binding induces dramatic alterations in the overall receptor structure. Interestingly, although specific agonist- and antagonist-induced conformational changes have been identified, both structures are incompatible with hsp interaction (Allan et al. 1992; Beekman et al. 1993). We have extended these studies to examine a wide variety of both ER and progesterone receptor (PR) modulators of different chemical derivation and have observed that the resulting conformational changes are independent of the ligand's chemical nature but rely instead solely on its agonist or antagonist activity. Cumulatively these results suggest, for ER and PR at least, that all agonists and antagonists promote dissociation of hsp and permit the interaction of receptor with specific DNA sequences in the regulatory regions of target gene promoters (Kumar and Chambon 1988; Martinez and Wahli 1989; Tsai et al. 1989). Consequently, it appears that steps downstream of DNA binding are likely responsible for distinguishing between agonist- and antagonist-activated receptors.

A great deal of attention has focused on the elucidation of the finals steps in the steroid hormone receptor transduction pathway. Although it has been shown that steroid receptors can interact directly with TFIIB, a constituent of the general transcription machinery, it is believed that their interaction with cell- or promoter-specific adaptors which facilitate indirect interactions with the general transcription machinery may be of equal importance (Ing et al. 1992; Webb et al. 1995). In the case of ER these studies have led to the identification of a protein (ERAP160) which specifically associates with hormone- but not antihormone-activated ER (Halachmi et al. 1994). It is possible that this protein could function as a transcriptional coactivator or adaptor, allowing the interaction of hormone- but not antihormone-activated ER with the general transcriptional machinery. However, determination of the precise role of ERAP160 in the ER signal transduction awaits its molecular cloning and analysis.

Another interesting protein TAF-II30b has been shown to physically interact with ER and modulate its activity in a reconstituted in vitro transcription system (Jacq et al. 1994). However, like ERAP160, its activity in a bona fide target cell is currently unknown. Several additional candidates for receptor coactivator proteins have been identified, raising the intriguing possibility that different receptors utilize different coactivators, or that the same receptor uses different coactivators under different circumstances. Based on these results we speculate that promoter- or cell-specific processes could be regulated by developing compounds which facilitate one type of interaction with the general transcription apparatus over another. We have pursued this avenue of investigation using ER as a model.

1.3 Identification of Cell-Specific ER Modulators

The biological actions of estradiol appear to be mediated through a single type of receptor which is differentially expressed in target cell nuclei. Although several variant ERs derived from the same gene have been identified, their biological role in regulating the cellular responses to estrogen have not yet been determined (Fuqua et al. 1993). Thus, the task of developing novel tissue selective ER modulators requires the identification of compounds that will interact with a single type of ER and modify its structure in such a way as to differentially regulate ER targets. The distinct biology manifested by estradiol and the steroidal antiestrogens ICI164,384 and ICI182,780 represents a very striking example of this concept (Hu et al. 1993; Wakeling et al. 1991). These antiestrogens, which differ only from estradiol in that they contain a 7α-alkylamide side chain, bind to the receptor and competitively inhibit agonist binding. In addition to this activity, however, these antiestrogens also manifest their effects by acting as pseudoagonists, mimicking some of the actions of estrogen, but leading ultimately to the commitment of ER to a transcriptionally nonproductive pathway. Since we have established that estradiol and the pure antihormones induce distinct structural alterations within ER and that both compounds promote delivery of the receptor to DNA, we believe that the cellular mechanisms which distinguish between agonists and antagonist must lie downstream of DNA binding (McDonnell et al. 1991, 1995). We propose further that it is the ability of the cell to distinguish between the estradiol- and ICI164,384-induced structural alterations within ER that leads to their unique biology (McDonnell et al. 1994, 1995). It is tempting to speculate that the distinct biological activity displayed by testosterone and dihydrotestosterone in males may relate to their ability to induce different structural alterations within the androgen receptor (AR) which can be distinguished by the transcriptional machinery in target cells.

The ER does not function independently on most target gene promoters but interacts with other promoter- or cell-specific factors to manifest its biology (McDonnell et al. 1992; Meyer et al. 1990; Tora et al. 1988; Tzukerman et al. 1994; Ylikomi et al. 1992). Even though under natural circumstances estrogen facilitates these specific interactions, it is likely that some, but not all, associations will be tolerant of structural perturbations in ER structure induced by synthetic agonists and antagonists. Based on this hypothesis, and the data presented above, we considered that compounds could be developed which would function as ER agonists in some cell contexts and as antagonists in others.

1.4 Specific Domains Within ER Interact with the Cellular Transcription Machinery and Mediate the Transcriptional Effects of Estradiol

The molecular mechanisms which transmit the hormonal signal from ER to the general transcription apparatus are slowly being elucidated. We and others have shown that ER contains two distinct regions required for maximal transcriptional activity (Berry et al. 1990; McDonnell et al. 1993; Tzukerman et al. 1994). A *trans*-activating function-1

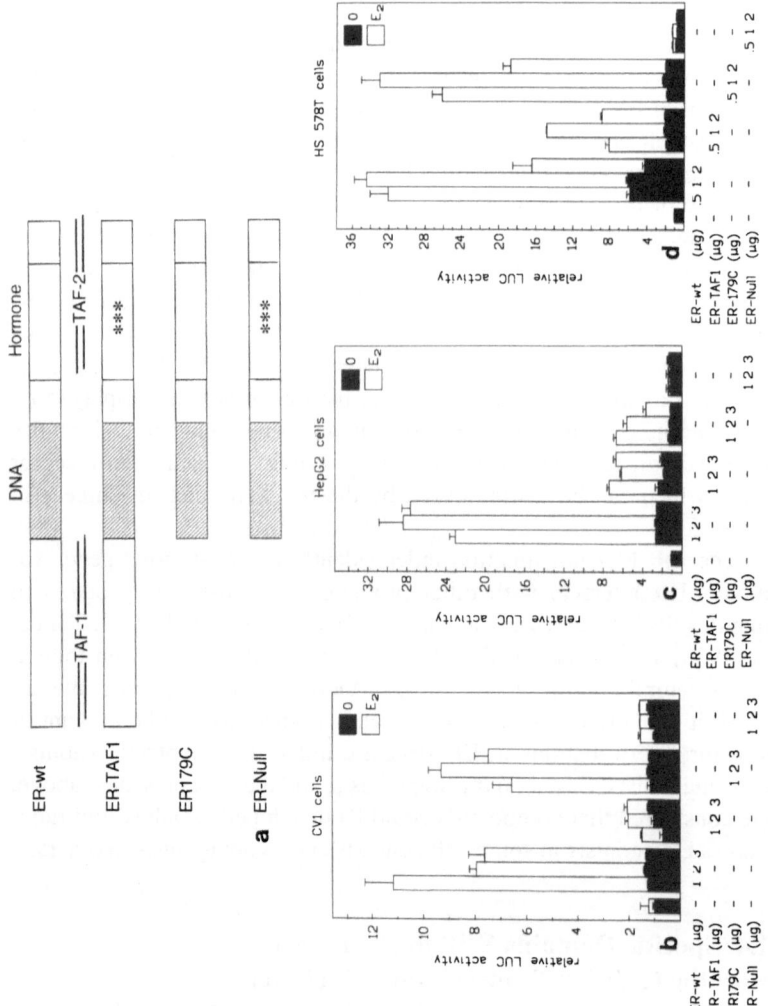

Fig. 1a–d. Legend see p. 7

(TAF-1) is located in the amino terminus of ER and a TAF-2 is located in the carboxyl terminus. The precise nature of these transactivation sequences is unknown at the current time. It is, however, possible to create specific receptor mutations which result in the removal of one or both of these TAFs. Using a series of mutant receptors, (Fig. 1a) we have shown that the transcriptional activity of TAF-1 and TAF-2 are dependent on the cell and promoter context (Tzukerman et al. 1994). When assayed on an ERE-thymidine kinase (ERE-TK) promoter we determined that in most cells both TAF-1 and TAF-2 were required for maximal transcriptional activity, but that each could manifest independent transcriptional activity (Fig. 1b–d). Specifically, we observed that in CV-1 cells, the ER-179C receptor, containing only TAF-2, exhibited about 80% the transcriptional activity of the wild-type receptor whereas TAF-1 was inactive (Fig. 1b). Similar results were obtained when these receptor mutants were assayed in the HS578T cell line (Fig. 1c), whereas in HepG2 cells maximal ER transcriptional activity required both activation sequences. We concluded from these experiments that TAF-1 and TAF-2 functioned in unique ways to permit ER-dependent transcription and that the cellular context played an important part in determining the efficacy of each transactivator.

To complete our analysis of ER transcriptional activity we performed a similar analysis of TAF-1 and TAF-2 function on additional promoters to evaluate the contribution of promoter context to their activity.

◀ **Fig. 1a–d.** The transcriptional activity of estrogen receptor (*ER*) is influenced by cell context. **a** Wild-type (*ER-wt*) and mutant *ERs* used in these experiments. The *asterisks* represent point mutations in the hormone-binding domain of ER where the amino acid residues at positions 538, 542, and 545 were replaced with alanines by site-directed mutagenesis. These mutations specifically disabled TAF-2. CV-1 (**b**) HepG2 (**c**), or HS578T (**d**) cells were transiently cotransfected with increasing concentrations of receptor expression vectors as indicated, together with an ERE-TK-Luciferase reporter plasmid and pRSV-β-GAL as an internal control for transfection efficiency. Cells were treated with or without estradiol for 36 h and assayed for luciferase and β-galactosidase activity. The relative luciferase activity was calculated by dividing the normalized luciferase value at a given point by that obtained in the absence of transfected receptor or ligand. The data shown indicate the mean SEM of triplicate estimations. [This figure is reprinted from *Molecular Endocrinology* (Tzukerman et al. 1994) with permission]

D. P. McDonnell et al.

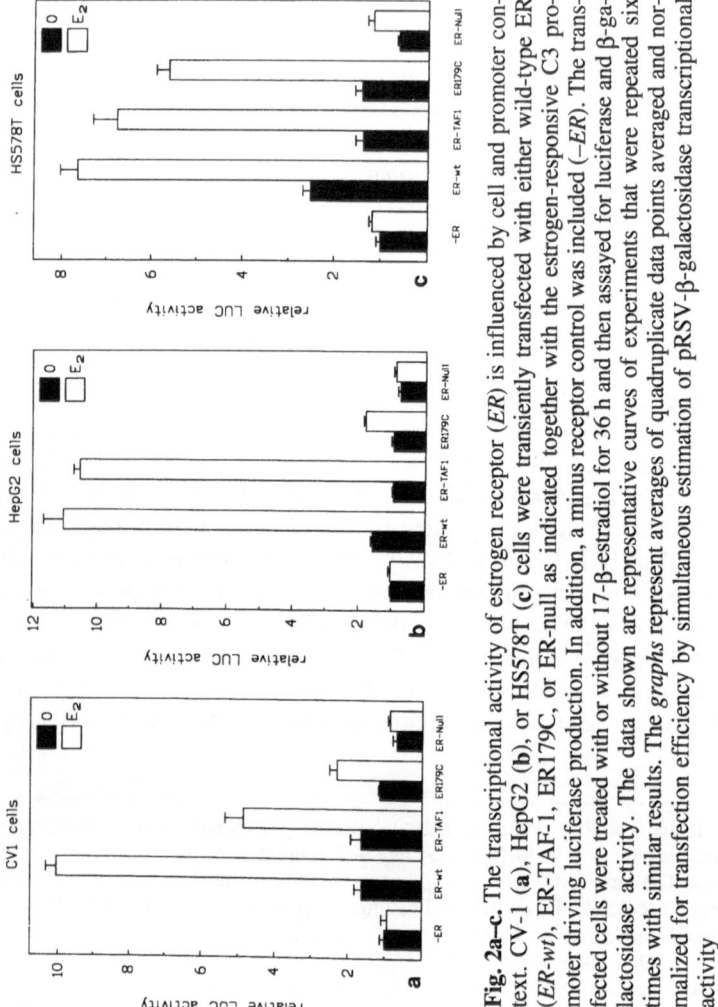

Fig. 2a–c. The transcriptional activity of estrogen receptor (*ER*) is influenced by cell and promoter context. CV-1 (**a**), HepG2 (**b**), or HS578T (**c**) cells were transiently transfected with either wild-type ER (*ER-wt*), ER-TAF-1, ER179C, or ER-null as indicated together with the estrogen-responsive C3 promoter driving luciferase production. In addition, a minus receptor control was included (*–ER*). The transfected cells were treated with or without 17-β-estradiol for 36 h and then assayed for luciferase and β-galactosidase activity. The data shown are representative curves of experiments that were repeated six times with similar results. The *graphs* represent averages of quadruplicate data points averaged and normalized for transfection efficiency by simultaneous estimation of pRSV-β-galactosidase transcriptional activity

The results of the most informative study, those obtained on the estrogen responsive complement 3 (C3) promoter, are shown in Fig. 2. In CV-1 cells, as noticed before, both TAF-1 and TAF-2 are required for maximal ER responsiveness on the C3 promoter. In striking contrast to the results obtained with the ERE-TK promoter (Fig. 1b), we observed that maximal estrogen responsiveness can occur in the absence of a functional TAF-2. Additionally, we find that TAF-2 alone can not activate transcription (Fig. 2b). Using the same promoter we find that in the breast cancer cell line HS578T both TAF-1 and TAF-2 can function independently to a level equivalent to wild-type ER (ER-wt) (Fig. 2c). The results obtained by surveying ER transcriptional activity on this and other reporters (not shown) suggest that TAF-1 and TAF-2 activity can be dramatically different when assayed in different cell and promoter contexts. Conceptually, therefore, it appears to us that compounds which differentially influence the synergistic interactions between TAF-1 and TAF-2 or those that influence the interaction of either TAF with other transcription factors have the potential to exhibit tissue (or pathway) selective ER modulatory activity.

1.5 The Pharmacology of ER Ligands Is Influenced by the Transcriptional Activity of TAF-1 and TAF-2

In order to define the role of the individual TAFs in determining the pharmacology of ER ligands, we assayed the activity of a series of known modulators in contexts where we had shown differences in the requirement for TAF-1 and TAF-2 activities. As demonstrated above, we have shown that on the ERE-TK promoter in CV-1 cells, TAF-2, but not TAF-1, can exhibit independent transcriptional activity. Therefore, in this context, it is likely that the biological activity of ER is manifest through TAF-2 within the intact receptor. This provided us with a unique opportunity to evaluate the role of this activation domain in determining the pharmacology of ER ligands. The results of this analysis are shown in Fig. 3. As expected, in this TAF-2-dependent context, estrogen functions as an agonist (Fig. 3a) and the pure antiestrogen ICI164,384 does not demonstrate any agonist activity. Interestingly, tamoxifen which functions as an ER partial agonist in some tissues, notably the uterus, does not induce ER transcriptional activity (Kedar et al. 1994). Not surprisingly therefore, both tamoxifen and ICI164,384 oppose the action of estrogen in this context when administered simultaneously (Fig. 3b). Thus, in contexts where TAF-2 activity is required for transcriptional activity, estradiol alone manifests agonist activity

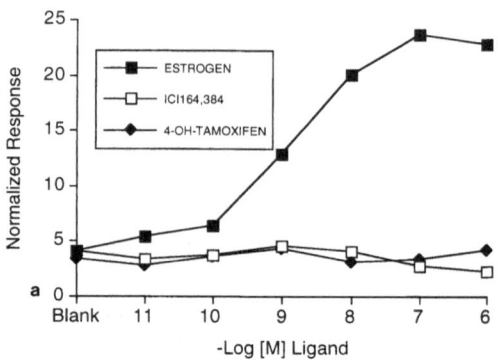

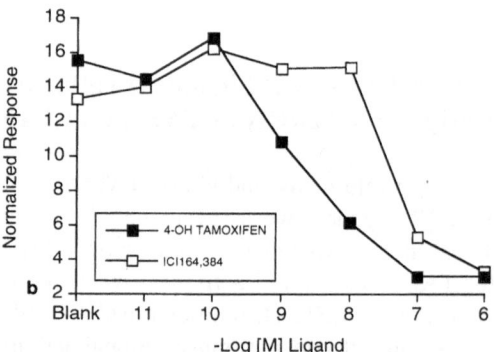

Fig. 3a,b. Tamoxifen functions as an antagonist in contexts where TAF-2 is required for estrogen receptor (ER) transcriptional activity. Monkey kidney CV-1 cells were transiently cotransfected with a human ER expression vector (ER-wt) together with an ERE-TK-Luciferase reporter plasmid and a pRSV-β-galactosidase expression vector (as an internal control for transfection efficiency). **a** The agonist activity of 17-β-estradiol, ICI164,384 or 4-OH-tamoxifen was evaluated by incubating the transfected cells with increasing concentrations of test compounds for 40 h. **b** Antagonist activities were assayed by incubating the transfected cells for 40 h in the presence of a constant amount of 17-β-estradiol (10^{-8} M) and an increasing concentration of either 4-OH-tamoxifen or ICI164,384 as indicated. Subsequently the transfected cells were harvested and assayed for luciferase and β-galactosidase activities. The normalized luciferase activity was calculated by dividing the raw luciferase ($\times 10^4$ units) for each point by the β-galactosidase activity [$(A_{415} \times 10^5)$/time in minutes)]. Each *data point* in this experiment represents the average of quadruplicate determinations of the transcriptional activity under a given experimental condition

whereas ICI164,384 and tamoxifen oppose this activity by functioning as TAF-2 antagonists.

We extended our analysis of ER by accessing its ligand responsiveness on the C3 promoter in HepG2 cells; a context where TAF-1 was the dominant transcriptional regulator. As shown above, TAF-2 is inactive on the C3 promoter whereas ER-TAF-1 alone can function as an independent transactivator (Fig. 2b). This unique system provided an assay with which to study the role of TAF-1 in determining the pharmacology of ER modulators. The results of this analysis are shown in Fig. 4. When assayed in the agonist mode, estrogen functioned as an efficient activator of ER transcriptional activity. Interestingly, however, whereas the pure antiestrogen ICI164,384 did not manifest any significant agonist activity, tamoxifen functioned as a partial agonist. This partial agonist activity of tamoxifen is in contrast to what we observed in a TAF-2-dependent context (Fig. 3a). Thus we concluded from these experiments that the partial agonist activity of tamoxifen was specifically related to its ability to function as an agonist in contexts where TAF-1 was the dominant activator.

The antagonist activities of these compounds were examined in the same context. As before, ICI164,384 functioned as a pure antagonist of estradiol activity, whereas tamoxifen only partially antagonized this activity. Thus, it is clear that one part of the antagonist action of tamoxifen is its ability to antagonize TAF-2 activity. However, since TAF-2 is not required in this cell and promoter context then tamoxifen functions as an inefficient antagonist.

These results are particularly interesting as they clearly distinguish the ER antagonist activity of ICI164,384 from that of tamoxifen. Specifically, we notice that ICI164,384 exhibits ER-antagonist action in a manner which is independent of TAF-1 or TAF-2 in a given cell or promoter context. Previous studies have suggested that ICI164,384 manifests its inhibitory activity on ER by promoting receptor turnover, nuclear translocation, and/or impeding receptor dimerization (Arbuckle et al. 1992; Dauvois et al. 1992, 1993; Fawell et al. 1990). Although these are likely to be involved, we have clearly shown that this compound also allows delivery of some receptor to its target DNA sequence; however, this receptor is unable to activate transcription (McDonnell et al. 1995). Because we have shown that ICI164,384 can inhibit ER transcriptional activity in both TAF-1- and TAF-2-dominant contexts

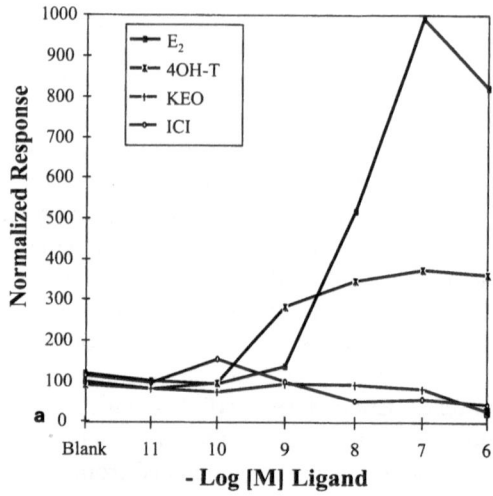

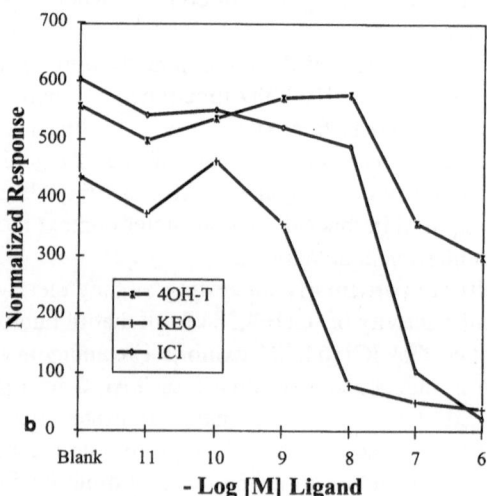

Fig. 4a,b. Legend see p. 13

we believe that interaction of ER with this compound modifies ER structure in such a way as to block the activities of both transactivators. In contrast, tamoxifen functions as an antagonist in a TAF-2 dominant background whereas in TAF-1 dependent contexts it manifests partial agonist activity. It is interesting however, that even in contexts where TAF-2 activity is not required, tamoxifen can function as a partial antagonist of estradiol activity. This suggests that in addition to inhibiting TAF-2 activity, in this context, tamoxifen has the ability to inhibit 50% of TAF-1 activity (Fig. 4b). Thus, we propose that tamoxifen, ICI164,384, and estradiol differentially affect ER structure in such a way as to modulate the way in which the TAFs are presented to the transcription apparatus. At one extreme is ICI164,384 which completely blocks TAF-1 and TAF-2 function and at the other is estradiol which can promote maximal activation of both TAFs. Tamoxifen represents an intermediate state where all TAF-2 activity is inhibited, but some TAF-1 activity remains. In support of this hypothesis we have been able to show, using a protease digestion assay, that all three compounds induce distinct structural alterations within ER (McDonnell et al. 1995). Cumulatively, these results provide a firm mechanistic basis for the partial agonist activity of tamoxifen and provide the framework with which to access the partial agonist activity of other modulators. They suggest also that it should be possible to develop additional compounds which by virtue of their activity on either TAF-1 or TAF-2 would exhibit cell and promoter-selective transcriptional activity.

◀ **Fig. 4a,b.** Tamoxifen functions as an agonist in contexts where the TAF-1 domain of estrogen receptor (ER) is transcriptionally active. HepG2 cells were transiently cotransfected with a human estrogen receptor expression vector (ER-wt) together with a C3-Luciferase reporter plasmid (complement 3 promoter) and a pRSV-β-galactosidase expression vector (as an internal control for transfection efficiency). **a** The agonist activity of 17-β-estradiol, ICI164,384, 4-OH-tamoxifen or keoxifene was evaluated by incubating the transfected cells with increasing concentrations of test compounds for 40 h.
b Antagonist activities were assayed by incubating the transfected cells for 40 h in the presence of a constant amount of 17-β-estradiol (10^{-8} M) and an increasing concentration of either 4-OH-tamoxifen or ICI164,384 or keoxifene, as indicated. Subsequently, the transfected cells were harvested and assayed for luciferase and β-galactosidase activities. The data were calculated as in Fig. 3

1.6 The Agonist Activity of Estradiol and Tamoxifen Can Be Distinguished by Specific ER Mutations

Thus far we have shown that the in cell contexts where TAF-1 is the dominant tranactivator estradiol can function as a full agonist while tamoxifen, at saturating levels of hormone, only induces about 30% of this activity (Fig. 4b). We were interested therefore in determining whether these agonist activities could be distinguished. In the HepG2 cell/C3 promoter context we showed that TAF-1 alone had independent transcriptional activity and that this was sufficient to permit estradiol to function as an agonist. Tamoxifen, on the other hand, functioned only as a partial agonist in this context. This data presented us with a paradox. Why in this cell context, in which TAF-2 is not required for estradiol action and TAF-1 is sufficient for maximal transcriptional activity, does tamoxifen function only as a partial agonist? One possibility is that, although TAF-2 activity may not be required, the distinct structural alterations within ER induced in this region by estradiol and tamoxifen may differentially affect the way TAF-1 is presented to the transcription apparatus. To address this possibility we assayed the transcriptional activity of a series of ER modulators on ER-wt and on a receptor mutant (ER-TAF-1) where TAF-2 activity had been destroyed by introducing three point mutations into the hormone-binding domain. These specific TAF-2 mutations were chosen as they had no measurable effects on the affinity or specificity of ER–ligand interactions, suggesting that these mutations had a minor effect on overall receptor structure (Fig. 5).

Fig. 5a–d. Specific mutations within TAF-2 facilitate the classification of estrogen receptor (*ER*) modulators. Human hepatocellular carcinoma cells (HepG2) were transiently transfected with either ER-wt, ER-TAF-1, or ER-null together with the estrogen-responsive complement 3 (C3) promoter fused to luciferase and a pRSV-β-galactosidase plasmid (as an internal control for transfection efficiency). The transfected cells were incubated for 36 h in the presence of ethanol carrier alone or in the presence of increasing concentrations of **a** 17-β-estradiol, **b** 4-OH-tamoxifen, **c** ICI164,384, or **d** keoxifene as indicated. Subsequently, the transfected cells were harvested and assayed for luciferase and β-galactosidase activities. The normalized luciferase activity was calculated by dividing the raw luciferase ($\times 10^4$ units) for each *point* by the β-galactosidase activity [($A_{415} \times 10^5$)/time in minutes)]. Each *panel* is representative of at least ten individual experiments

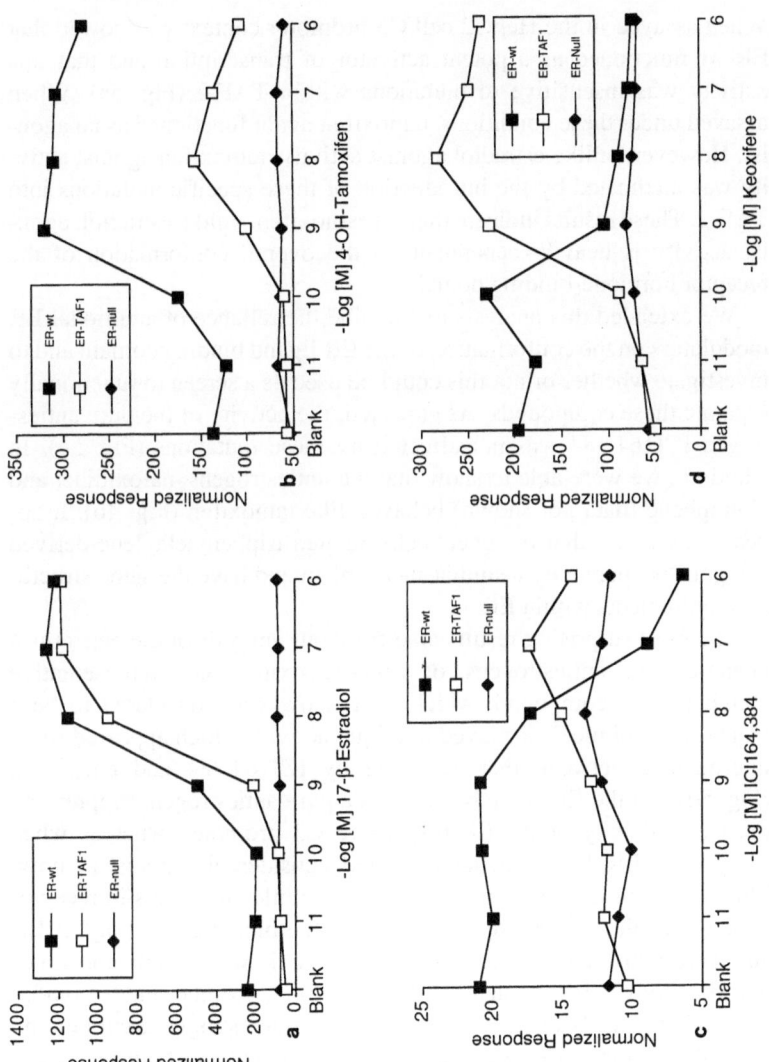

Fig. 5a–d. Legend see p. 14

When assayed in the HepG2 cell/C3 promoter context we showed that ER-wt functioned as a potent activator of transcription and that this activity was insensitive to mutations within TAF-2 (Fig. 5a). When assayed under these conditions, tamoxifen again functioned as an agonist. However, unlike estradiol agonist activity, tamoxifen agonist activity was attenuated by the introduction of these specific mutations into TAF-2. These results indicate that for tamoxifen, unlike estradiol, agonist activity is heavily dependent on the overall conformation of the receptor hormone-binding domain.

We extended this analysis to examine the reliance of additional ER modulators on the conformation of the ER ligand binding domain and to investigate whether or not this could be used as a screen to functionally separate these compounds. As expected, the activity of the pure antiestrogen ICI164,384 was not affected by these mutations (Fig. 5c). In addition , we were able to show that the antiestrogens, nafoxidine, and clomiphene (data not shown) behaved like tamoxifen (Fig. 4b). It appears, therefore, that the chemically related triphenylethylene-derived antagonists operate by a similar mechanism and have the same structural requirements within ER.

The most surprising result came from our analysis of the benzothiophene-derived antiestrogens, of which keoxifene is a representative member. This compound, which up to now was considered to be a tamoxifen "mimic," displayed a unique activity which appeared to be intermediate between that exhibited by ICI164,384 and tamoxifen (Fig. 5d). On the ER-wt, it behaved as a pure antiestrogen, suppressing the basal activity of the ER-responsive C3 promoter, whereas when assayed on ER-TAF-1 it demonstrated considerable agonist activity. This important observation indicated that subtle alterations in receptor structure could have profound effects on the biological activity of ER and its responsiveness to ligands. In addition, however, it demonstrates in a very convincing manner that ICI164,384, keoxifene, and tamoxifen are mechanistically distinct ER modulators. Interestingly, accumulating clinical information suggests that unlike ICI164, 384 both keoxifene and tamoxifen display partial agonist activity in vivo. (Black et al. 1994; Draper et al. 1993; Kedar et al. 1994; Love et al. 1992). However, preliminary studies indicate that the agonist activity of keoxifene is more restricted than that of tamoxifen. These data firmly support the hypothesis that tissue- or process-selective estrogenic activity can be obtained by developing compounds which induce specific structural alterations within ER. Additionally, it suggests that in vitro assays which are amenable to high throughput screens can be used to predict the likely clinical activity of an ER modulator.

1.7 Identification of Cellular Factors that Interact with and Modify the Biological Activity of the TAFs Within ER

The definition of two distinct regions within ER required for its productive interaction with the cellular transcription apparatus prompted us and others to identify the factors that mediate the interaction of ER with the general transcription machinery. As demonstrated in Fig. 1b, when assayed on minimal promoters such as the ERE-TK promoter, we observed that both TAF-1 and TAF-2 were required for maximal ER responsiveness. On more complex promoters this requirement varies according to the specific cell and promoter context in which the receptor activation sequence is assayed (Tzukerman et al. 1994). We have made similar observations in *Saccharomyces cerevisiae* where we have found that the human estrogen functions as a ligand-dependent transcription factor (Pham et al. 1991, 1992). Similar to what we had observed on minimal promoters in mammalian cells, neither TAF-1 nor TAF-2 could function independently when assayed in yeast. Thus, in view of the conserved nature of the transcription machinery from yeast to humans (Buratowski et al. 1989; Cavallini et al. 1988; Horikoshi et al. 1989; Poon and Weil 1993; Ranish et al. 1992; Ruppert et al. 1993), we considered that the identification of factors (or processes) which affect the ER TAFs in yeast may be enlightening with respect to their function in mammalian cells (Pham et al. 1992; Tzukerman et al. 1994). We were particularly interested in defining the role of TAF-2. Consequently, we introduced a construct expressing TAF-1 alone into yeast and screened for cellular mutations that would mimic TAF-2 function, allowing TAF-1 to function as an independent transactivator. Using this approach we identified a repressor protein, SSN6, which when mutated increases the transcriptional efficacy of TAF-1 to a level comparable to estradiol-activated ER-wt (Fig. 6) (McDonnell et al. 1992). Interestingly, in an *ssn6* mutant background we observe that the amount of estradiol required for half maximal activation of transcription by ER-wt decreases dramati-

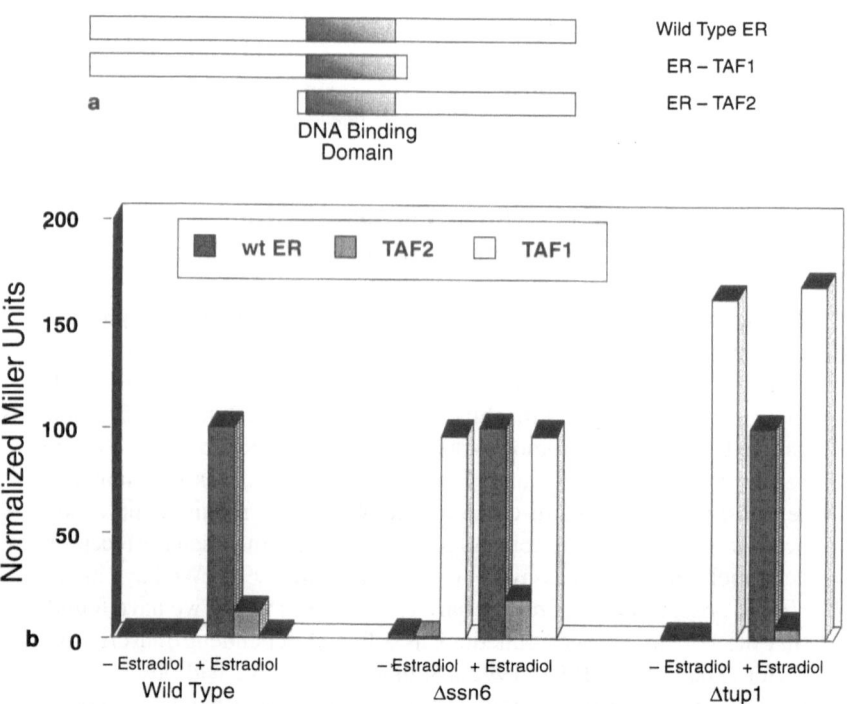

Fig. 6a,b. The TAF-1 function of estrogen receptor (*ER*) is regulated by the SSN6/TUP1 complex when assayed in *Saccharomyces cerevisiae*. **a** The constructs used in this experiment to express ER-wt, ER-TAF-1, or ER-TAF-2 in either wild-type yeast or in yeast strains bearing null mutations of *ssn6* or *tup1*. The *shaded region* represents the DNA binding domain of ER. **b** The transcriptional activity of ER-wt, TAF-1, or TAF-2 were assayed in wild-type yeast or in strains bearing null mutations in SSN6 or TUP1. These null mutants, Δ*ssn6* and Δ*tup1*, were created in the background of a yeast strain BJ5409 (matα, *ura3-52, trp1, leu2Δ1*) by insertional disruption of the coding sequences of these genes using a HIS3 or LEU2 selectable marker, respectively. The genotype was confirmed by Southern analysis. Transcriptional activity in these strains was accessed on an ERE-CYCl-β-galactosidase reporter in the presence or absence of 10^{-6} *M* β-estradiol. All values were normalized to the hormone activated ER-wt within each experiment, and represent the average of triplicate assays

cally from 10^{-9} M to 10^{-13} M (McDonnell et al. 1992). In addition, the antiestrogens nafoxidine, tamoxifen, and ICI164,384 function also as ER agonists in this cellular background, whereas in wild-type yeast they exhibit minimal agonist activity (McDonnell et al. 1992). It is known that the SSN6 protein functions in yeast as a transcriptional repressor of several different promoters (Cooper et al. 1994; Keleher et al. 1992). Recent biochemical and genetic evidence suggests that SSN6 works in concert with another protein, TUP1, to mediate its biological actions (Tzamarias and Struhl 1994; Williams and Trumbly 1991).

To examine whether a similar regulatory complex operated on ER, we created null mutations both of the *ssn6* and *tup1* genes and assayed the activity of ER and the individual TAFs in this background. The results of this analysis are shown in Fig. 6. The transcriptional activity of ER and its individual TAFs in a *tup1* knockout strain of yeast are similar to that observed in an *ssn6* knockout (Fig. 6). Specifically, in wild-type yeast, TAF-1 has no significant transcriptional activity, whereas in either an *ssn6* or *tup1* null background its activity is equivalent to that of the hormone-induced wild-type receptor. The effects of SSN6 and TUP1 on ER appear to be specific for TAF-1 as the transcriptional activity of TAF-2 is only marginally increased when assayed in an *ssn6* background. This suggests that both SSN6 and TUP1 are likely to be involved in repressing TAF-1 and ER-wt transcriptional activity. The identification of a cellular mutation of this type which can bypass the need for TAF-2 suggests possibly that the function of this activator in mammalian cells is to overcome the repressive effect of a homologue of this protein allowing TAF-1 to manifest transcriptional activity.

A considerable amount of attention of late has focused on the mechanism by which the SSN6/TUP1 complex represses transcription in yeast (Keleher et al. 1992). Several models have been proposed, as a consequence of this work, which may extrapolate to the regulation of ER function. One suggestion is that the SSN6/TUP1 functions as a general repressor of transcription and that specificity is afforded by specific activators which can overcome the transcriptional inhibitory action of chromatin structure on transcription (Cooper et al. 1994). If this were so, then it would appear that neither TAF-1 nor TAF-2 alone could accomplish this task in the presence of the SSN6/TUP1 repressor. However, their combined action in the context of the wild-type receptor could overcome the repressive activity of SSN6/TUP1. This model does not require a physical interaction of ER with the SSN6/TUP1 proteins. An alternative hypothesis has emerged from the studies of Herschbach et al. (1994) who demonstrated using a chromatin-free in vitro repression assay that the SSN6/TUP1 complex likely functions by blocking access of transcriptional activators to the general transcriptional machinery. Interestingly, the SSN6/TUP1 complex does not bind DNA directly but is tethered to target promoters through its association with other DNA-bound transcription factors with. In yeast, for instance, the SSN6/TUP1 complex has been shown to be involved in mating-type suppression where it represses transcription following its interaction with Mcm1 and $\alpha 2$ proteins at a-specific gene operators (Herschbach et al. 1994). Likewise, its activity as a mediator of glucose repression results as as consequence of its ability to interact with the MIG1 protein at responsive promoters. The recent work by Treitel and Carlson (1995) on the mechanism of the MIG/SSN6/TUP1 complex suggests that there may be parallels between the actions of SSN6/TUP1 on MIG1 and their actions on ER. MIG1 binds to specific sequences within the promoter of glucose-repressed genes. In this way it tethers SSN6/TUP1 to DNA and permits repression of target gene transcription. When MIG1 activity was assayed in a *ssn6* null background it functioned as an efficient transcriptional activator. Thus, MIG1 is a mediatior of glucose repression in some circumstances, although in instances when SSN6 activity is attenuated it can function as an activator. This is similar to what has been proposed for the thyroid hormone receptor, which functions as a repressor in the absence of ligand and as an activator when ligand is added (Chin 1994). Although the switch which converts the MIG1/SSN6/TUP1 complex from a repressor to an activator is unknown, it has been shown that MIG1 is differentially phosphorylated under different growth conditions. It is possible that this modification could affect MIG1/SSN6/TUP1 interactions and its ability to repress or potentiate transcription. Extrapolating to ER action, it is possible that SSN6/TUP1 functions as a repressor of ER action in the absence of hormone and that conformational changes (and/or phosphorylation changes) induced by hormone disrupts the interaction of the receptor with the SSN6/TUP1 repressor.

Currently, we are using a combined approach in yeast and mammalian cells to define the mechanism by which the SSN6/TUP1 repressor acts on ER. In addition, we have initiated a search for proteins (or processes) in mammalian cells which operate in a manner similar to SSN6/TUP1. Clearly, if such a protein(s) does exist, it would play a major role in defining cell-specific action of ER modulators.

1.8 A Model Describing the Pharmacology of ER Modulators

Considering published data and that generated in our own laboratory we believe that the majority of currently described antiestrogens efficiently deliver ER to DNA (Kumar and Chambon 1988; McDonnell et al. 1991). The distinct conformations of ER induced by receptor agonists and antagonists and additional constraints placed on these structures by the ERE (and adjacent sequences) influences the manner in which the cellular transcription machinery recognizes ligand-activated ER and ultimately determines the pharmacological activity of a given ligand (Allan et al. 1992). A model for these receptor/transcription–apparatus interactions is shown in Fig. 7. In this model we consider that receptor agonists bind to their cognate receptors, inducing a structure that allows the TAF-2 function within the hormone-binding domain to overcome the repressive effects of a complex comprising the mammalian homologues of the yeast SSN6 and TUP1 proteins. The function of this repressor complex is to prevent the interaction of the TAF-1 region with the general transcription apparatus either directly or indirectly via an adaptor protein. Upon displacement of the repressor complex by TAF-2, TAF-1 can interact with the putative adaptor molecule altering its conformation and allowing it to interact productively with the cellular transcription machinery. As observed in our studies (described above) there are some circumstances in which TAF-1 or TAF-2 can function independently. We propose that in contexts where TAF-1 alone is active that other transcription factors acting at the same promoter can overcome the inhibitory actions of the SSN6/TUP1 complex. In those cases where repressor activity is removed, TAF-2 activity is no longer required and so TAF-2 antagonists such as tamoxifen have the potential to exhibit partial agonist activity. The differences in ER structure induced by estradiol and tamoxifen probably account for the differences in their efficacy. We propose that the overall structure of the tamoxifen–ER complex only permits a weak interaction of the TAF-1s with the tran-

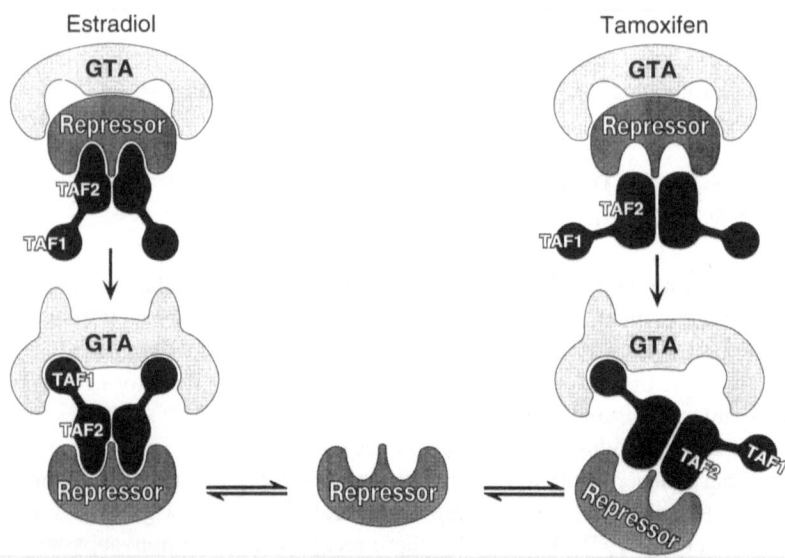

Fig. 7. Cellular discrimination between agonist- and antagonist-activated estrogen receptor (*ER*). Interaction of ER with either agonists or antagonists promotes the displacement of heat-shock proteins, facilitating dimerization and high-affinity DNA binding (McDonnell et al. 1991, 1995). We believe that hormone- and antihormone-activated ER are recognized differently by the cellular transcription machinery. Information from our work in yeast suggests that interaction of ER with agonists induces a structural alteration within ER which allows the TAF-2 domain to overcome the repressive effects of the SSN6/TUP1 complex (or an equivalent mammalian repressor). This allows a productive interaction of TAF-1 with the general transcription apparatus (*GTA*), possibly mediated through an adaptor protein. When occupied by a receptor antagonist we believe that the conformation of the ER is such that it cannot overcome the negative influence of the repressor. In addition, however, in cell contexts where other transcription factors lead to the displacement of the repressor the conformation of the TAF-1 domains are such that they interact weakly with the GTA, permitting partial agonist activity. This model provides a molecular explanation for the partial agonist activities of the triphenylethylene-derived antiestrogens such as tamoxifen and suggests that it should be possible to develop ligands which display different degrees of partial agonist efficacy

scriptional machinery. In the case of pure antiestrogens such as ICI164,384, the conformation of ER is such that it cannot displace the SSN6–TUP1 repressor complex. In addition, even when the repressor is displaced by other proteins the TAF-1 and TAF-2 within the ICI164,384/ER complex are not accessible to the transcription apparatus.

This model reflects our current understanding of ER action, providing a conceptual explanation for the cell-specific partial agonist activities exhibited by some ER antagonists. However, it should be considered as a working model directing future experimental strategies.

1.9 Final Comments

Considerable progress has been made in understanding the mechanism of action of ER and how it modulates important biological processes in breast cancer cells. In particular, the discovery of novel mechanisms for ER activation which occur in the absence of ligand suggests that ER may be a key point of convergence of multiple cellular signaling pathways (Ignar-Trowbridge et al. 1992, 1993; Smith et al. 1993). It will be important to determine whether these different mechanisms of activation are responsible for the regulation of distinct subsets of ER-responsive genes.

Another important advance in this field is the observation that the cellular and promoter context and the composition of the estrogen-responsive sequences within a given target gene are critical in determining the efficacy of ER modulators (Dana et al. 1994; Tzukerman et al. 1994). This information may lead to the development of ER modulators which display tissue, cellular, or promoter specificity. It is envisaged that compounds will be developed which will function as antagonists of a series of biological processes in one cell but would display favorable agonist activities in another cell.

Finally, the recent advances in our understanding of ER action has provided some clues with respect to the molecular basis for cellular resistance to antihormone therapy observed in the clinic. However, it is the application of this new knowledge to the discovery of novel therapeutics and to improvements in the way we diagnose and treat estrogen-dependent cancers that will justify the efforts spent to understand ER action.

Acknowledgments. Some of the work presented was supported by an NIH grant (DK48897) to DPM and by a research grant from Ligand Pharmaceuticals Inc.

References

Allan GF, Leng X, Tsai S-T, Weigel NL, Edwards DP, Tsai M-J, O'Malley BW (1992) Hormone and antihormone induce distinct conformational changes which are central to steroid receptor activation. J Biol Chem 267:19513–19520

Arbuckle ND, Dauvois S, Parker MG (1992) Effects of antioestrogens on the DNA binding activity of oestrogen receptor in vitro. Nucleic Acids Res 20:3849–3844

Beato M (1989) Gene regulation by steroid hormones. Cell 56:335–344

Beekman JM, Allan GF, Tsai SY, Tsai M-J, O'Malley BW (1993) Transcriptional activation by the estrogen receptor requires a conformational change in the ligand binding domain. Mol Endocrinol 7:1266–1274

Berry M, Metzger D, Chambon P (1990) Role of the two activating domains of the oestrogen receptor in the cell-type and promoter-context dependent agonistic activity of the anti-oestrogen 4-hydroxytamoxifen. EMBO J 9:2811–2818

Black LJ, Sato M, Rowley ER, Magee DE, Bekele A, Williams DC, Cullinan GJ, Bendele R, Kaufman FR, Bensch WR, Frolik CA, Termine JD, Bryant HU (1994) Raloxifene (LY 139481) prevents bone loss and reduces serum cholesterol without causing uterine hypertrophy in ovaiectomized rats. J Clin Invest 93:63–93

Buratowski S, Hahn S, Sharp PA, Guarante L (1989) Function of a yeast TATA element-binding protein in a mammalian transcription system. Nature 334:37–42

Cavallini B, Huet J, Plassat J-L, Sentenac A, Egly J-M, Chambon P (1988) A yeast activity can substitute for the HeLa cell TATA box factor. Nature 334:77–80

Chin WW (1994) Molecular mechanisms of thyroid hormone action. Thyroid 4:389–393

Clark JH, Peck EJ (1979) Female sex steroids: receptors and function. Monogr Endocrinol 14:4–36

Cooper JP, Roth SY, Simpson RT (1994) The global transcriptional regulators, SSN6 and TUP1, play distinct roles in the establishment of a repressive chromatin structure. Genes Dev 8:1400–1410

Dana SL, Hoener PA, Wheeler DL, Lawrence CL, McDonnell DP (1994) Novel estrogen response elements identified by genetic selection in yeast are differentially responsive to estrogens and antiestrogens in mammalian cells. Mol Endocrinol 8:1193–1207

Dauvois S, Danielian PS, White R, Parker MG (1992) Antiestrogen ICI 164,384 reduces cellular estrogen receptor content by increasing its turnover. Proc Natl Acad Sci USA 89:4037–4041

Dauvois S, White R, Parker MG (1993) The antiestrogen ICI182780 disrupts estrogen receptor nucleoplasmic shuttling. J Cell Sci 106:1377–88

Draper MW, Flowers DE, Huster WJ, Neild JA (1993) Effects of raloxifene (LY 139481 HCL) on biochemical markers of bone and lipid metabolism in healthy post-menopausal women. Handelstrykkeriet, Aalborg

Fawell SE, White R, Hoare S, Sydenham M, Page M, Parker MG (1990) Inhibition of estrogen receptor-DNA binding by the "pure" antiestrogen ICI 164,384 appears to be mediated by impaired receptor dimerization. Proc Natl Acad Sci USA 87:6883–6887

Fuqua SAW, Chamness GC, McGuire WL (1993) Estrogen receptor mutations in breast cancer. J Cell Biochem 51:135–139

Halachmi S, Marden E, Martin G, MacKay H, Abbondanza C, Brown M (1994) Estrogen receptor-associated proteins: possible mediators of hormone-induced transcription. Science 264:1455–1458

Henderson BE, Ross RK, Pike MC (1993) Hormonal chemoprevention of cancer in women. Science 259:633–638

Herschbach BM, Arnaud MB, Johnson AD (1994) Transcriptional repression directed by the yeast alpha 2 protein in vitro. Nature 370:309–311

Horikoshi M, Wang CK, Fujii H, Cromlish JA, Weil PA, Roeder RG (1989) Cloning and structure of a yeast gene encoding a general transcription initiation factor TFIID that binds to the TATA box. Nature 341:299–303

Hu HF, Veroni M, DeLuise M, Wakeling A, Sutherland R, Watts CK, Zalcberg JR (1993) Circumvention of tamoxifen resistance by the pure anti-estrogen ICI182,780. Int J Cancer 55:873–876

Ignar-Trowbridge DM, Nelson KG, Bidwell MC, Curtis SW, Washburn TF, McLachlan JA, Korach KS (1992) Coupling of dual signalling pathways: epidermal growth factor action involves the estrogen receptor. Proc Natl Acad Sci USA 89:4658–4662

Ignar-Trowbridge DM, Teng CT, Ross KA, Parker MG, Korach KS, McLachlan JA (1993) Peptide growth factors elicit estrogen receptor dependent activation of an estrogen-responsive element. Mol Endocrinol 7:992–998

Ing NH, Beekman JM, Tsai SY, Tsai MJ, O'Malley BW (1992) Members of the steroid hormone receptor superfamily interact with TFIIB (S300-II). J Biol Chem 267:17617–17623

Jacq X, Brou C, Lutz Y, Davidson I, Chambon P, Tora L (1994) Human TAFII30 is present in a distinct TFIID complex and is required for transcriptional activation by the estrogen receptor. Cell 79:107–117

Kedar RP, Bourne TH, Powles TJ, Collins WP, Ashley SE, Cosgrove DO, Cambell S (1994) Effects of tamoxifen on uterus and ovaries of postmenopausal women in a randomised breast cancer prevention trial. Lancet 343:1318–1321

Keleher CA, Redd MJ, Schultz J, Carlson M, Johnson AD (1992) Ssn6-Tup1 is a general repressor of transcription in yeast. Cell 68:709–719

Kettel LM, Murphy AA, Mortola JF, Liu JH, Ulmann A, Yen SS (1991) Endocrine responses to long-term administration of the anti-progesterone RU486 in patients with pelvic endometriosis. Fertil Steril 56:402–407

Kumar V, Chambon P (1988) The estrogen receptor binds tightly to its responsive element as a ligand-induced homodimer. Cell 55:145–156

Love RR, Mazess RB, Barden HS, Epstein S, Newcomb PA, Jordan VC, Carbone PP, DeMets DL (1992) Effects of tamoxifen on bone mineral density in postmenopausal women with breast cancer. N Engl J Med 326:852–856

Martinez E, Wahli W (1989) Cooperative binding of estrogen receptor to imperfect estrogen-responsive DNA elements correlates with their synergistic hormone-dependent enhancer activity. EMBO J 8:3781–3791

McDonnell DP, Nawaz Z, O'Malley BW (1991) In situ distinction between steroid receptor binding and transactivation at a target gene. Mol Cell Biol 11:4350–4355

McDonnell DP, Vegeto E, O'Malley BW (1992) Identification of a negative regulatory function for steroid receptors. Proc Natl Acad Sci USA 89:10563–10567

McDonnell D, Vegeto E, Gleeson MAG (1993) Nuclear hormone receptors as targets for new drug discovery. Biotechnology 11:1256–1261

McDonnell DP, Clemm DL, Imhof MO (1994) Definition of the cellular mechanisms which distinguish between hormone and antihormone activated steroid receptors. Semin Cancer Biol 5:503–513

McDonnell DP, Clemm DL, Herman T, Goldman ME, Pike JW (1995) Analysis of estrogen receptor function in vitro reveals three distinct classes of antiestrogens. Mol Endocrinol 9:659–669

Meyer ME, Pornon A, Ji J, Bocquel MT, Chambon P, Gronemeyer H (1990) Agonist and antagonist properties of RU486 on the functions of the human progesterone receptor. EMBO J 9:3923–3932

Pham TP, Hwung Y-P, McDonnell DP, O'Malley BW (1991) Transactivation
    functions facilitate the disruption of chromatin structure by estrogen recep-
    tor derivatives in vivo. J Biol Chem 266:18179–18187
Pham TP, Hwung Y-P, Santiso-Mere D, McDonnell DP, O'Malley BW (1992)
    Ligand-dependent and independent functions of the transactivation regions
    of the human estrogen receptor in yeast. Mol Endocrinol 6:1043–1050
Poon D, Weil PA (1993) Immunopurification of yeast TATA-binding protein
    and associated factors: presence of transcription factor IIIB transcriptional
    activity. J Biol Chem 268:15325–15328
Pratt WB, Hutchinson KA, Scherrer LW (1992) Steroid receptor folding by
    heat-shock proteins and composition of the receptor heterocomplex. Trends
    Endocrinol Metab 3:326–333
Ranish JA, Lane WS, Hahn S (1992) Isolation of two genes that encode sub-
    units of the yeast transcription factor II. Science 255:1127–1130
Ruppert S, Wang EH, Tjian R (1993) Cloning and expression of human
    TAFII250: a TBP-associated factor implicated in cell-cycle regulation. Na-
    ture 362:175–179
Sanchez ER, Hirst M, Scherrer LC, Tang HY, Welsh MJ, Harmon JM, Sim-
    mons SSJ, Ringold GM, Pratt WB (1990) Hormone free mouse glucocorti-
    coid receptors overexpressed in Chinese hamster ovary cells are localized
    to the nucleus and are associated with both hsp70 and hsp90. J. Biol. Chem.
    265:20123–20130
Smith CL, Conneely OM, O'Malley BW (1993) Modulation of the ligand-in-
    dependent activation of the human estrogen receptor by hormone and anti-
    hormone. Proc Natl Acad Sci USA 90:6120–6124
Smith DF, Toft DO (1993) Steroid receptors and their associated proteins. Mol
    Endocrinol 7:4–11
Sunderland MC, Osborne CK (1991) Tamoxifen in pre-menopausal patients
    with metatastic breast cancer: a review. J Clin Oncol 9:1283–1297
Tora L, Gronemeyer H, Turcotte B, Gaub M-P, Chambon P (1988) The N-ter-
    minal region of the chicken progesterone receptor specifies target gene acti-
    vation. Nature 333:185–188
Treitel MA, Carlson M (1995) Repression by SSN6-TUP1 is directed by
    MIG1, a repressor/activator protein. Proc Natl Acad Sci USA 92:3132–
    3136
Tsai S-Y, Tsai M-J, O'Malley BW (1989) Cooperative binding of steroid hor-
    mone receptors contibutes to transcriptional synergism at target enhancer
    elements. Cell 57:443–448
Tzamarias D, Struhl K (1994) Functional dissection of the yeast Cyc8-Tup1
    transcriptional co-repressor complex. Nature 369:758–761
Tzukerman MT, Esty A, Santiso-Mere D, Danielian P, Parker MG, Stein RB,
    Pike JW, McDonnell DP (1994) Human estrogen receptor transcriptional

capacity is determined by both cellular and promoter context and mediated by two functionally distinct intramolecular regions. Mol Endocrinol 8:21–30

Wakeling AE, Dukes M, Bowler J (1991) A potent specific pure antiestrogen with clinical potential. Cancer Res 51:3867–3873

Webb P, Lopez GN, Uht RM, Kushner PJ (1995) Tamoxifen activation of the estrogen receptor/AP-1 pathway: potential origin for the cell-specific estrogen-like effects of antiestrogens. Mol Endocrinol 9:443–456

Williams FE, Trumbly RJ (1991) The CYC8 and TUP1 proteins involved in glucose repression in Saccharomyces cerevisiae are associated in a protein complex. Mol Cell Biol 11:3307–3316

Ylikomi T, Bocquel MT, Berry M, Gronemeyer H, Chambon P (1992) Cooperation of proto-signals for nuclear accumulation of estrogen and progesterone receptors. EMBO J 11:3681–3694

2 Estrogen–Receptor and Antiestrogen–Receptor Complexes: Cell- and Promoter-Specific Effects and Interactions with Second Messenger Signaling Pathways

B. S. Katzenellenbogen, M. M. Montano, W. L. Kraus,
S. M. Aronica, N. Fujimoto, and P. LeGoff

2.1   Introduction and Overview 29
2.2   Estrogen and Antiestrogen Binding and Discrimination by the ER   31
2.3   The Carboxy-Terminal F Domain of the ER:Role in the
        Transcriptional Activity of the Receptor and the Effectiveness
        of Antiestrogens as Estrogen Antagonists in Different Target Cells 35
2.4   Cross-Talk Between the ER and Second Messenger
        Signaling Pathways in Cells 36
2.5   Phosphorylation of the ER 41
2.6   Antiestrogen Selectivity and Promoter Dependence in the
        cAMP-Dependent Signaling Pathway Involvement in Activation
        of the Transcriptional Activity of ERs Occupied by Antiestrogens   42
2.7   Bidirectional Cross-Talk Between Estrogen and cAMP
        Signaling Pathways 43
References 45

2.1 Introduction and Overview

Estrogens regulate the differentiation, growth, and functioning of many reproductive tissues. They also exert important actions on other tissues, including bone, liver, and the cardiovascular system. Most of the actions of estrogens appear to be exerted via the estrogen receptor (ER) of target cells, an intracellular receptor that is a member of a large superfamily of proteins that function as ligand-activated transcription factors, regulating the synthesis of specific RNAs and proteins. Because of this diversity of estrogen target tissues, much current interest focuses on trying to understand the basis for the cell and promoter context-dependent actions of estrogens and antiestrogens. Antiestrogens, which antagonize the actions of estrogens, have much potential as important therapeutic agents. Although antiestrogens bind to the ER in a manner that is competitive with estrogen, they fail to effectively activate gene transcription (Jordan and Murphy 1990; Katzenellenbogen et al. 1985; Santen et al. 1990). Two of the major challenges in studies on antiestrogens are to understand what accounts for their antagonistic effectiveness, as well as the partial agonistic effects of some antiestrogens, and to understand how one can attain tissue-selective agonist/antagonist effects of these compounds.

In order to address these issues, many of our analyses have focused in detail on the hormone-binding domain of the ER, regions E and F, since this domain of the receptor contains both hormone-binding and hormone-dependent transactivation functions of the receptor. In our attempts to understand how the ER discriminates between estrogen and antiestrogen ligands, and between different categories of antiestrogens, we have generated and analyzed variant human ERs with mutations in the ER hormone-binding domain and studied the activity of these receptors on different estrogen-responsive genes in several cell backgrounds when liganded with antiestrogenic or estrogenic ligands. These studies, and those of others, have provided consistent evidence for the promoter-specific and cell-specific actions of the estrogen-occupied and antiestrogen-occupied ER. In addition, although the binding of estrogens and antiestrogens is mutually competitive, studies with ER mutants indicate that some of the contact sites of estrogens and antiestrogens are likely different. Our recent studies reveal that the presence of the C-terminal F domain of the ER is important in the transcription activation and repression activities of antiestrogens and that it affects the magnitude of liganded ER bioactivity in a cell-specific manner (Montano et al. 1994, 1995). The influence of the F domain on the agonist/antagonist balance and potency of antiestrogens supports its specific modulatory role in the ligand-dependent interaction of ER with components of the transcription complex. These studies (Ince et al. 1993, 1995; Katzenellenbogen et al. 1993; Pakdel and Katzenellenbogen 1992; Wrenn and Katzenellenbogen 1993, see below) have provided evidence for a regional dissociation of the hormone-binding and transcription-activation regions in domain E of the receptor and have also shown that mutations in the hormone-binding domain and deletions of C-terminal regions result in ligand discrimination mutants, that is, receptors that are differentially altered in their ability to bind and/or mediate the actions of estrogens versus antiestrogens.

In addition, in studies described below, we have observed that protein kinase activators enhance the transcriptional activity of the ER and alter the agonist/antagonist balance of some antiestrogens, suggesting that changes in the cellular phosphorylation state should be important in determining the effectiveness of antiestrogens as estrogen antagonists. The ability of estrogens and antiestrogens to also increase cAMP levels in target cells suggests that the interaction of estrogens with second messenger signaling pathways may be bidirectional.

2.2 Estrogen and Antiestrogen Binding and Discrimination by the ER

A variety of studies (Berry et al. 1990; Fujimoto and Katzenellenbogen 1994; Pakdel et al. 1993a, b; Pakdel and Katzenellenbogen 1992; Reese and Katzenellenbogen 1991, 1992a, b; Tzukerman et al. 1994; Wrenn and Katzenellenbogen 1993) have provided strong documentation that the response of genes to estrogen and antiestrogen depend on several important factors: (a) the nature of the ER, i.e., whether it is wild type or variant; (b) the ligand; (c) the promoter; and (d) the cell context. The gene response, in addition, can be modulated by cAMP, growth factors, and agents that affect protein kinases and cell phosphorylation (Aronica and Katzenellenbogen 1991, 1993; Cho et al. 1994; Fujimoto and Katzenellenbogen 1994; Kraus et al. 1993). These factors, no doubt, account for differences in the relative agonism/antagonism of antiestrogens, for instance, tamoxifen, on different genes and in different target cells such as those in breast cancer cells versus uterine or bone cells.

Although both estrogens and antiestrogens bind within the hormone-binding domain, the association must differ because estrogen binding activates a transcriptional enhancement function, whereas antiestrogens fully or partially fail in this role. Antiestrogens are believed to act in large measure by competing for binding to the ER and altering the conformation of the ER such that the receptor fails to effectively activate gene transcription. In addition, antiestrogens exert anti-growth factor activities via a mechanism that requires ER but is still not fully understood (Freiss and Vignon 1994). Models of antiestrogen action at the molecular level are beginning to emerge, and recent biological studies also indicate that antiestrogens fall into at least two distinct categories: antiestrogens such as tamoxifen that are mixed or partial agonists/antagonists (type I), and compounds such as ICI164,384 that are complete/pure antagonists (type II). The type I antihormone–ER complexes appear to bind as dimers to EREs; there, they block hormone-dependent transcription activation mediated by region E of the receptor, but are believed to have little or no effect on the hormone-independent transcription activation function located in region A/B of the receptor (Berry et al. 1990). Thus, they are generally partial or mixed agonist/antagonists, and their action must involve some subtle difference in ligand–receptor interaction very likely associated with the basic or polar side chain that characterizes the antagonist members of this class. In the case of the more complete antagonists such as ICI164,384, ER conformation must clearly differ from that of the estrogen-occupied ER since alteration in ER binding to DNA and reduction of the ER content of target cells appear to contribute to (Dauvois et al. 1992; Fawell et al. 1990b), but may not fully explain, the pure antagonist character of this antiestrogen (Reese and Katzenellenbogen 1992b). Of note is the fact that antiestrogens, whether steroidal or nonsteroidal, typically have a bulky side chain which is basic or polar. This side chain is important for antiestrogenic activity; removal of this side chain results in a compound which is no longer an antiestrogen and, instead, has only estrogenic activity. Therefore, interaction of this side chain with the ER must play an important role in the interpretation of the ligand as an antiestrogen.

In order to understand how the ER "sees" an antiestrogen as different from an estrogen, we have used site-directed and random chemical mutagenesis of the ER cDNA to generate ERs with selected changes in the hormone-binding domain. We have been particularly interested in identifying residues in the hormone-binding domain important for the

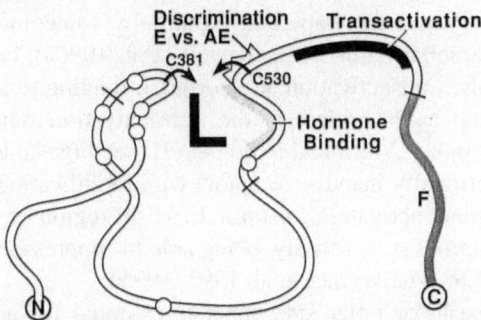

Fig. 1. "Map" of functions in the human estrogen receptor hormone-binding domain (HBD). Domain E, amino acids 302-553, is shown as is the very C-terminal domain *F*, amino acids 554-595. Some regions considered to be important in hormone binding, discrimination between estrogen (*E*) and antiestrogen (*AE*), and transactivation are highlighted. The ligand (*L*) is portrayed in a region representing the ligand binding pocket of the receptor. *Open circles* indicate amino acids in the HBD where our mutational analyses have shown mutational changes to affect the affinity or stability of hormone binding. See text for description. *N*, N-terminus; *C*, C-terminus of receptor

ligand binding of estrogen and antiestrogen and transactivation functions of the receptor and in elucidating the mechanism by which the ER differently interprets agonistic and antagonistic ligands. Our studies have indicated that selective changes near amino acid 380 and amino acids 520-530 and changes at the C-terminus of the ER result in ER ligand discrimination mutants (Montano et al. 1994, 1995; Pakdel et al. 1993b; Pakdel and Katzenellenbogen 1992). These data provide evidence that some contact sites of the receptor with estrogen and antiestrogen differ and that the conformation of the receptor with estrogen and antiestrogen must also be different as a consequence (Danielian et al. 1993; Pakdel and Katzenellenbogen 1992, references therein).

Our structure–function analysis of the hormone-binding domain of the human ER has utilized region-specific mutagenesis of the ER cDNA and phenotypic screening in yeast, followed by the analysis of interesting receptor mutants in mammalian cells (Katzenellenbogen et al. 1993; Wrenn and Katzenellenbogen 1993). A great advantage of the yeast system is that it allows the rapid screening of a library of many mutants, a situation that is not possible in mammalian cells. Our observations, as well as very important studies by Malcolm Parker and colleagues (Danielian et al. 1992, Fawell et al. 1990a), have shown a separation of the transactivation and hormone-binding functions of the ER, with amino acids critical in the transactivation function of the receptor being more C-terminal in domain E (see Fig. 1). Interestingly, some transcriptionally inactive receptors with modifications in this domain E C-terminal activation function 2 (AF-2) region of the ER have potent dominant negative activity, being able to suppress the activity of the wild-type ER in cells (Ince et al. 1993, 1995).

Since the basic or polar side chain is essential for antiestrogenic activity, and our previous studies identified cysteine 530 as the amino acid covalently labeled by affinity-labeling ligands (Harlow et al. 1989), we introduced by site-directed mutagenesis of the ER cDNA changes of specific charged residues close to C530 (Pakdel and Katzenellenbogen 1992). Interestingly, two mutants in which lysines at position 529 and 531 were changed to glutamines, so that the local charge was changed, resulted in ligand discrimination mutants. These receptors showed an approximately thirtyfold increased potency of antiestrogen in suppressing estradiol (E$_2$)-stimulated reporter gene activity. Interestingly, these mutant receptors had a reduced binding affinity for estrogens, but retained unaltered binding affinity for antiestrogen. These observations suggest that we are able to differentially alter estrogen and antiestrogen effectiveness by rather modest changes in the ER, and that the region near C530 is a critical one for sensing the fit of the side chain of the estrogen antagonist. Studies from the Parker Laboratory (Danielian et al. 1993) have shown that nearby residues (i.e., G525 and M521 and/or S522 in the mouse ER) are also importantly involved in conferring differential sensitivity to these two categories of ligands.

2.3 The Carboxy-Terminal F Domain of the ER:
Role in the Transcriptional Activity of the Receptor and
the Effectiveness of Antiestrogens as Estrogen Antagonists
in Different Target Cells

Among the nuclear hormone receptors, ER is unusual in having a large C-terminal F region (42 amino acids) and its role in ER bioactivity and in the activities of antiestrogens has not been well defined. Previous studies by us and others have shown that domain F is not required for transcriptional response to E_2 (Kumar et al. 1986, 1987; Pakdel et al. 1993a), and, in addition, our studies have shown that this region does not affect the turnover rate of ER in target cells (Pakdel et al. 1993a). However, a more complete examination (Montano et al. 1994, 1995) has revealed that the presence of the F domain is important in the transcription activation and repression activities of antiestrogens, and that it affects the magnitude of liganded ER bioactivity in a cell-specific manner. Thus, in ER-negative breast cancer cells and Chinese hamster ovary (CHO) cells, E_2 stimulated equally the transcription of several estrogen-responsive promoter-reporter gene constructs with wild-type ER, and with ER lacking the carboxy-terminal F domain or ER altered in the F domain by point mutations. Of interest, the antiestrogens trans-hydroxytamoxifen and ICI164,384, which showed considerable agonistic activity on some of the reporter constructs with the wild-type ER, showed no agonistic activity with the ER lacking the F domain. In addition, the antiestrogens were more potent antagonists of E_2-stimulated transcription by the F domain-deleted ER than by wild-type ER. Interestingly, the effect of the F domain was very dependent on the cell type. In HeLa human cervical cancer cells, the F domain-deleted ER exposed to E_2 was much less effective than wild-type ER in stimulating transcription, and antiestrogens were less potent in suppressing E_2-stimulated transcription by the F domain-deleted ER. Since we find that the F domain does not appear to affect estrogen or antiestrogen ligand-binding affinity or DNA binding of the receptor, the fact that this region makes the liganded ER either more or less transcriptionally effective in different cells suggests that it plays an important role in maintaining ER conformation optimal for protein–protein interactions needed for transcriptional effectiveness. Its influence on the agonist/antagonist balance and potency of antiestrogens further supports its specific modulatory role in the ligand-dependent interaction of ER with components of the transcription complex. Therefore, these studies (Montano et al. 1994, 1995) as well as several others by us (Cho and Katzenellenbogen 1993; Fujimoto and Katzenellenbogen 1994; Katzenellenbogen et al. 1995) and those of others have highlighted that beyond differences in estrogen and antiestrogen binding to the ER, the cell context and promoter context of the ER influences the estrogenic/antiestrogenic activity of antiestrogens.

2.4 Cross-Talk Between the ER and Second Messenger Signaling Pathways in Cells

Evidence for cross-talk between steroid hormone receptors and signal transduction pathways has been increasing. Expression of AP-1, a transcription factor of the Fos/Jun heterodimer known to mediate the protein kinase-C (PKC) pathway (Angel et al. 1987), was shown to suppress steroid hormone receptor-mediated gene expression (Schüle et al. 1990), most likely through direct protein–protein interaction between steroid receptors and these oncoproteins (Yang-Yen et al. 1990). In addition, the ovalbumin gene promoter containing a half-palindromic estrogen-responsive element (ERE) was coactivated by ER and Fos/Jun oncoproteins (Gaub et al. 1990). Thus, interaction between these oncoproteins and steroid hormone receptors resulted in cell-specific inhibitory or stimulatory effects on transcriptional activation (Gaub et al. 1990; Shemshedini et al. 1991; Strähle et al. 1988; Yang-Yen et al. 1990).

Previous studies by us and others (Aronica and Katzenellenbogen 1991; Katzenellenbogen and Norman 1990; Sumida et al. 1988; Sumida and Pasqualini 1989, 1990) documented upregulation of intracellular progesterone receptor, an estrogen-stimulated protein, by insulin-like growth factor-I (IGF-I), epidermal growth factor, phorbol ester, and cAMP in MCF-7 human breast cancer cells and uterine cells. The fact that the stimulation by these diverse agents was blocked by antiestrogen suggested that these agents were presumably acting through the ER pathway (Aronica and Katzenellenbogen 1991; Katzenellenbogen and Norman 1990; Sumida et al. 1988; Sumida and Pasqualini 1989). In addition, the fact that protein kinase inhibitors also blocked the effects of estrogen, cAMP, and growth factors suggested the involvement of phosphorylation in these responses. We therefore undertook studies to examine directly whether activators of protein kinases can modulate transcriptional activity of the ER.

In primary cultures of uterine cells using transient transfection experiments with simple estrogen-responsive reporter genes, we examined the ability of these agents to stimulate ER-mediated gene transcription and also compared the ability of these multiple agents to alter the phosphorylation state of the endogenous uterine ER protein. The results of our study (Aronica and Katzenellenbogen 1993) indicate that E_2, IGF-I, and agents which raise intracellular cAMP are able to stimulate ER-mediated transactivation and ER phosphorylation. The fact that antiestrogen (ICI164,384) evokes a similar increase in ER phosphorylation without a similar increase in transcription activation indicates that an increase in overall ER phosphorylation does not necessarily result in increased transcriptional activity. Also, the observation that transcriptional activation by the ER was nearly completely suppressed by the protein kinase inhibitors H8 and PKI, while the increase in phosphorylation was reduced by 50%–75%, indicates that the correlation between transcriptional activation and overall ER phosphorylation is not direct but does suggest that some of the effects of E_2, IGF-I and cAMP on ER-regulated transactivation are mediated through the activity of protein kinases. Our findings, demonstrating a clear effect of these agents on ER-mediated transactivation, suggest that these agents might also regulate endogenous estrogen target genes, such as that encoding the progesterone receptor, by similar cellular mechanisms.

In order to examine some of the molecular mechanisms controlling transcription of the progesterone receptor gene, we cloned the rat progesterone receptor gene 5'-region and identified two functionally distinct promoters (Kraus et al. 1993). The two distinct promoters in the rat progesterone receptor gene exhibited differential responsiveness to E_2 and to ER-dependent stimulation by cAMP. The functional differences between these two promoters may lead to altered expression of the A and B progesterone receptor isoforms and, thereby, influence cellular responsiveness to progestins (Kraus et al. 1993).

In MCF-7 human breast cancer cells and other cells, we found that activators of PKA and PKC markedly synergize with E_2 in ER-mediated

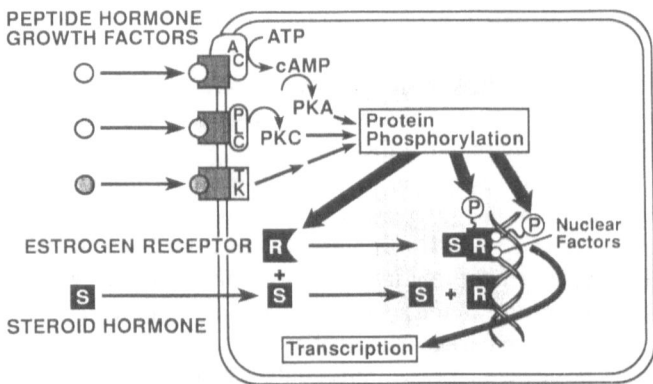

Fig. 2. Protein kinase–estrogen receptor transcriptional synergism. See text for description. *AC*, adenylate cyclase; *PLC*, phospholipase C; *TK*, tyrosine kinase; *PKC*, protein kinase C; *PKA*, protein kinase A; *R*, estrogen receptor; *S*, steroid hormone; *P*, phosphate groups on proteins

transcriptional activation and that this transcriptional synergism shows cell and promoter specificity (Cho and Katzenellenbogen 1993; Fujimoto and Katzenellenbogen 1994; Kraus et al. 1993). The synergistic stimulation of ER-mediated transcription by E_2 and protein kinase activators did not appear to result from changes in ER content or in the binding affinity of ER for ligand or the ERE DNA, but rather may be a consequence of a stabilization or facilitation of interaction of target components of the transcriptional machinery, possibly either through changes in phosphorylation of ER or other proteins important in ER-mediated transcriptional activation (Cho and Katzenellenbogen 1993).

   Figure 2 shows a model indicating how we think the protein kinase-ER transcriptional synergism might occur. Agents influencing protein kinase pathways may enhance intracellular protein phosphorylation resulting in either phosphorylation of the ER itself or the phosphorylation of nuclear factors with which the receptor interacts in mediating transcription. Likewise, there is evidence that the steroid hormone itself can alter receptor conformation increasing its susceptibility to serve as a substrate for protein kinases (Ali et al. 1993; Aronica and Katzenellenbogen 1993; Denton et al. 1992; Lahooti et al. 1994; Le Goff et al. 1994). Therefore, agents which increase phosphorylation may, either

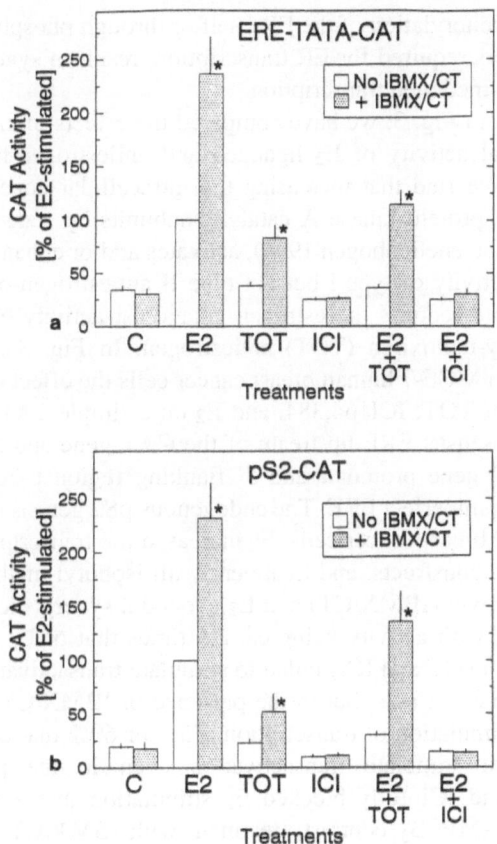

Fig. 3a,b. Effect of IBMX/CT on the ability of E_2 and antiestrogens to stimulate transactivation of ERE-TATA-CAT (**a**) and pS2-CAT (**b**), and on the ability of antiestrogens to suppress E_2-stimulated transactivation. MCF-7 cells were transfected with the indicated reporter plasmid and an internal control plasmid that expresses β-galactosidase and were treated with the agents indicated for 24 h. Each *bar* represents the mean ± SEM (*n*=3 experiments). * indicates significant difference from the no IBMX/CT cells ($p < 0.05$). *C*, control ethanol vehicle; E_2, 10^{-9} M; *TOT, trans*-hydroxytamoxifen (10^{-6} M); *ICI*, ICI 164384 (10^{-6} M); *IBMX*, 3-isobutyl-1-methyl-xanthine (10^{-4} M); *CT*, cholera toxin (1 μg/ml). (From Fujimoto and Katzenellenbogen 1994)

through phosphorylation of the ER itself or through phosphorylation of nuclear factors required for ER transcription, result in synergistic activation of ER-mediated transcription.

As shown in Fig. 3, we have compared the effects of cAMP on the transcriptional activity of E_2 liganded and antiestrogen-liganded ER complexes. We find that increasing the intracellular concentration of cAMP, or of protein kinase A catalytic subunits by transfection (Fujimoto and Katzenellenbogen 1994), activates and/or enhances the transcriptional activity of type I but not type II antiestrogen-occupied ER complexes and reduces the estrogen antagonist activity of the type I trans-hydroxy-tamoxifen (TOT) antiestrogen. In Fig. 3a,b, we have determined in MCF-7 human breast cancer cells the effect of cAMP on the activity of TOT, ICI164,384, and E_2 on a simple TATA promoter with one consensus ERE upstream of the CAT gene and on the more complex pS2 gene promoter and 5' flanking region (–3000 to +10) containing an imperfect ERE. The endogenous pS2 gene is regulated by E_2 in MCF-7 breast cancer cells. E_2 increased the transcription of both of these gene constructs, and treatment with isobutyl-methyl-xanthine and cholera toxin (IBMX/CT) and E_2 evoked a synergistic increase in transcription, with activity being ca. 2.5 times that of E_2 alone. Both antiestrogens (TOT and ICI) failed to stimulate transactivation of these reporter gene constructs, but in the presence of IBMX/CT, TOT gave significant stimulation of transcription (85% or 60% that of E_2 alone). ICI failed to stimulate transactivation even in the presence of IBMX/CT, and ICI fully blocked E_2 stimulation in the presence or absence of cAMP. By contrast, treatment with IBMX/CT reduced the ability of TOT to inhibit E_2 transactivation. While TOT returned E_2 stimulation down to that of the control in the absence of IBMX/CT (compare open bars E_2 vs. E_2+TOT in Fig. 3), TOT only partially suppressed the E_2 stimulation in the presence of IBMX/CT (compare stippled bars E_2 vs. E_2+TOT in Fig. 3).

Although alteration in the agonist and antagonist activity of TOT was observed with promoter–reporter constructs containing a simple TATA promoter and a more complex pS2 promoter, elevation of cAMP did not enhance the transcription by either TOT or E2 of the reporter plasmid ERE-thymidine kinase-CAT (Fujimoto and Katzenellenbogen 1994). Thus, this phenomenon is promoter specific. Of note, cAMP and PKA catalytic subunit transfection failed to evoke transcription by the more pure antiestrogen ICI164,384 with any of the promoter–reporter constructs tested. These findings, which document that stimulation of the PKA signaling pathway activates the agonist activity of tamoxifen-like antiestrogens, may in part explain the development of tamoxifen resistance by some ER-containing breast cancers. They also suggest that the use of antiestrogens such as ICI164,384 that fail to activate ER transcription in the presence of cAMP may prove more effective for long-term antiestrogen therapy in breast cancer.

2.5 Phosphorylation of the ER

Since our data suggested that estrogens, and agents that activate protein kinases, might influence ER transcription by altering the state of phosphorylation of the ER and/or other factors required for ER regulation of transcription, we undertook studies to examine directly the effects of these agents on ER phosphorylation. In addition, we compared the effects of the type I and type II antiestrogens on phosphorylation of the ER. E2, the antiestrogens, *trans*-hydroxy-tamoxifen and ICI164,384, as well as PKA and PKC activators enhanced overall ER phosphorylation, and in all cases, this phosphorylation appeared to be on serine residues (Le Goff et al. 1994).

Tryptic phosphopeptide patterns of wild-type and domain A/B-deleted receptors and site-directed mutagenesis of several serines involved in known protein kinase consensus sequences allowed us to identify serine 104 and/or serine 106 and serine 118, all three being part of a serine-proline motif, as major ER phosphorylation sites. Mutation of these serines to alanines so as to eliminate the possibility of their phosphorylation resulted in an approximately 40% reduction in transactivation activity in response to E2 while mutation of only one of these serines showed an approximately 15% decrease in activation (Le Goff et al. 1994). Of note, E2 and antiestrogen-occupied ERs showed virtually identical two-dimensional phosphopeptide patterns, suggesting similar sites of phosphorylation. In contrast, the cAMP-stimulated phosphorylation likely occurs on different phosphorylation sites as indicated by some of our mutational studies (Le Goff et al. 1994) and this aspect remains under investigation in our laboratory.

Related studies in COS-1 cells by the Chambon laboratory (Ali et al. 1993) also identified serine 118 as being a major estrogen-regulated phosphorylation site. In MCF-7 cells, the Notides laboratory has also identified S118 as a site of ER phosphorylation but has observed S167 to be the most prominent site of phosphorylation in these cells (Arnold et al. 1994). Aurrichio and coworkers (Castoria et al. 1993) have also provided strong evidence for ER phosphorylation on tyrosine 537. The roles of these phosphorylations in the activities (transcriptional and other) of the ER remains an area of great interest.

2.6 Antiestrogen Selectivity and Promoter Dependence in the cAMP-Dependent Signaling Pathway Involvement in Activation of the Transcriptional Activity of ERs Occupied by Antiestrogens

Our data provide strong evidence for the involvement of cAMP-dependent signaling pathways in the agonist actions of tamoxifen-like estrogen antagonists. The promoter specificity of the transcriptional enhancement phenomenon suggests that factors in addition to ER are probably being modulated by PKA pathway stimulation. The findings imply that changes in the cAMP content of cells, which can result in activation of the agonist activity of tamoxifen-like antiestrogens, might account, at least in part, for the resistance to antiestrogen therapy that is observed in some breast cancer patients. Of interest, MCF-7 cells transplanted into nude mice fail to grow with tamoxifen treatment initially, but some hormone-resistant cells grow out into tumors after several months of tamoxifen exposure (Gottardis and Jordan 1988; Jordan and Murphy 1990; Osborne et al. 1991). Studies have shown that this resistance to tamoxifen is, more correctly, a reflection of tamoxifen stimulation of proliferation, representing a change in the interpretation of the tamoxifen–ER complex and its agonist/antagonist balance. It is of interest that we found the pS2 gene, which is under estrogen and antiestrogen regulation in breast cancer (Brown et al. 1984), to be activated by tamoxifen in the presence of elevated cAMP. By contrast, however, antiestrogens such as ICI164,384, shown in many systems to be more complete estrogen antagonists, are not changed in their agonist/antagonist balance by increasing intracellular concentrations of cAMP. Therefore,

ICI164,384-like compounds may prove to be more efficacious and less likely to result in antiestrogen-stimulated growth.

The transcriptional enhancement we have observed between PKA activators and ER occupied by tamoxifen-like antiestrogens and E_2 provides further evidence for cross-talk between the ER and signal transduction pathways regulated by cAMP that are important in ER-dependent responses.

2.7 Bidirectional Cross-Talk Between Estrogen and cAMP Signaling Pathways

For many years, steroid hormones and peptide hormones have been considered to act via distinctly different mechanisms, the former via intracellular receptors acting through the genome and the latter via membrane-localized receptors that initially affect extranuclear activities, including the generation of second messengers such as cAMP. However, there has been increasing evidence for interactions between cAMP and estrogen in enhancing the growth of the mammary gland and breast cancer cells (Sheffield and Welsch 1985; Silberstein et al. 1984) and for cAMP induction of estrogen-like uterine growth (see Aronica et al. 1994, references therein). As early as 1967, Szego and Davis (1967) demonstrated a very rapid, acute elevation of uterine cAMP by estrogen treatment of rats in vivo that was confirmed in other reports, but several subsequent studies either failed to confirm this observation or reported only minimal effects that were considered to represent indirect effects of estrogen on cAMP mediated by estrogen-induced release of uterine epinephrine (see Aronica et al. 1994, references therein). Recently, cAMP and other protein kinase activators have been documented to synergize with steroid hormone-occupied receptors, leading to enhanced receptor-mediated transcription (Aronica and Katzenellenbogen 1993; Beck et al. 1993; Cho and Katzenellenbogen 1993; Fujimoto and Katzenellenbogen 1994; Groul and Altschmeid 1993; Sartorius et al. 1993), possibly by a mechanism involving phosphorylation of the receptor or associated transcription factors (Ali et al. 1993; Aronica and Katzenellenbogen 1993; Le Goff et al. 1994; Montminy et al. 1986).

Our recent studies have shown that estrogen activates adenylate cyclase, markedly increasing the concentration of cAMP in estrogen-responsive breast cancer and uterine cells in culture and in the intact uterus of rats treated with estrogen in vivo, in a manner that does not require new RNA or protein synthesis (Aronica et al. 1994). The intra-cellular concentrations of cAMP achieved by low, physiological levels of estrogen are substantial and sufficient to stimulate cAMP response element (CRE)-mediated gene transcription. Therefore, this non-genomic action of the steroid hormone estrogen involves the production of an important second messenger and the resultant activation of second messenger-stimulated genes. These findings document a two-way direc-tionality in the cross-talk between steroid hormone- and cAMP-signal-ing pathways.

Although the hormonal specificity in the stimulation of cAMP is consistent with it being mediated by a high-affinity, ER-like binder, the nature of the potential membrane binder remains to be determined. Likewise, the mechanism by which estrogen enhances intracellular cAMP levels remains to be further examined. Several other publications have indicated an important role for sex hormone-binding globulin in the actions of sex steroids in enhancing intracellular cAMP (Fissore et al. 1994, Fortunati et al. 1993). The role of serum factors, including sex hormone-binding globulin, remains an important aspect for future ana-lyses. These current observations imply a possibly broad involvement of steroid hormone action on cyclic nucleotide and second-messenger sys-tems. The increasing evidence for rapid, membrane effects of estrogens and progestins and for vitamin D in a variety of target cells suggests that this aspect of steroid hormone action deserves further investigation (Ke and Ramirez 1987, 1990; Kim and Ramirez 1986; Lieberherr et al. 1993; Pappas et al. 1995).

Thus, data from this laboratory and others provide increasing evi-dence for extensive two-way cross-talk between estrogen and cAMP signaling pathways: in one way, cAMP is able to enhance the transcrip-tion of estrogen-regulated genes containing EREs; in the second way, estrogens appear able to act via the cAMP system to potentially regulate cAMP-mediated gene transcription. Further analyses of the underlying mechanisms in a variety of target cells should provide further insights in understanding the biology and regulation of estrogen-responsive cells.

References

Ali S, Metzger D, Bornert J-M, Chambon P (1993) Modulation of transcriptional activation by ligand-dependent phosphorylation of the human oestrogen receptor A/B region. EMBO J 12:1153–1160

Angel P, Imagawa M, Chiu M, Stein B, Imbra RJ, Rahmsdorf HJ, Jonat C, Herrlich P, Karin M (1987) Phorbol ester-inducible genes contain a common cis-acting element recognized by a TPA-modulator trans-acting factor. Cell 49:729–739

Arnold SF, Obourn JD, Jaffe H, Notides AC (1994) Serine 167 is the major estradiol-induced phosphorylation site on the human estrogen receptor. Mol Endocrinol 8:1208–1214

Aronica SM, Katzenellenbogen BS (1991) Progesterone receptor regulation in uterine cells: stimulation by estrogen, cyclic adenosine 3',5'-monophosphate and insulin-like growth factor I and suppression by antiestrogens and protein kinase inhibitors. Endocrinology 128:2045–2052

Aronica SM, Katzenellenbogen BS (1993) Stimulation of estrogen receptor-mediated transcription and alteration in the phosphorylation state of the rat uterine estrogen receptor by estrogen, cyclic adenosine monophosphate, and insulin-like growth factor-I. Mol Endocrinol 7:743–752

Aronica SM, Kraus WL, Katzenellenbogen BS (1994) Estrogen action via the cAMP signaling pathway: stimulation of adenylate cyclase and cAMP-regulated gene transcription. Proc Natl Acad Sci USA 91:8517–8521

Beck CA, Weigel NL, Moyer ML, Nordeen SK, Edwards DP (1993) The progesterone antagonist RU486 acquires agonist activity upon stimulation of cAMP signaling pathways. Proc Natl Sci Acad USA 90:4441–4445

Berry M, Metzger D, Chambon P (1990) Role of the two activating domains of the oestrogen receptor in the cell-type and promoter-context dependent agonistic activity of the anti-oestrogen 4-hydroxytamoxifen. EMBO J 9:2811–2818

Brown AMC, Jeltsch JM, Roberts M, Chambon P (1984) Activation of pS2 gene transcription is a primary response to estrogen in the human breast cancer cell line MCF-7. Proc Natl Acad Sci USA 81:6344–6348

Castoria G, Migliaccio A, Green S, Di Domenico M, Chambon P, Auricchio F (1993) Properties of a purified estradiol-dependent calf uterus tyrosine kinase. Biochemistry 32:1740–1750

Cho H, Katzenellenbogen BS (1993) Synergistic activation of estrogen receptor-mediated transcription by estradiol and protein kinase activators. Mol Endocrinol 7:441–452

Cho H, Aronica SM, Katzenellenbogen BS (1994) Regulation of progesterone receptor gene expression in MCF-7 breast cancer cells: a comparison of the effects of cyclic AMP, estradiol, IGF-1 and serum factors. Endocrinology 134:658–664

Danielian PS, White R, Lees JA, Parker MG (1992) Identification of a conserved region required for hormone dependent transcriptional activation by steroid hormone receptors. EMBO J 11:1025–1033

Danielian PS, White R, Hoare SA, Fawell SF, Parker MG (1993) Identification of residues in the estrogen receptor that confer differential sensitivity to estrogen and hydroxytamoxifen. Mol Endocrinol 7:232–240

Dauvois S, Danielian PS, White R, Parker MG (1992) Antiestrogen ICI164,384 reduces cellular estrogen receptor content by increasing its turnover. Proc Natl Acad Sci USA 89:4037–4041

Denton RR, Koszewski NJ, Notides AC (1992) Estrogen receptor phosphorylation: hormonal dependence and consequence on specific DNA binding. J Biol Chem 267:7263–7268

Fawell SE, Lees JA, White R, Parker MG (1990a) Characterization and colocalization of steroid binding and dimerization activities in the mouse estrogen receptor. Cell 60:953–962

Fawell SE, White R, Hoare S, Sydenham M, Page M, Parker MG (1990b) Inhibition of estrogen receptor-DNA binding by the "pure" antiestrogen ICI164,384 appears to be mediated by impaired receptor dimerization. Proc Natl Acad Sci USA 87:6883–6887

Fissore F, Fortunati N, Comba A, Fazzari A, Gaidano G, Berta L, Frairia R (1994) The receptor-mediated action of sex steroid binding protein (SBP, SHBG): accumulation of cAMP in MCF-7 cells under SBP and estradiol treatment. Steroids 59:661–667

Fortunati N, Fissore F, Fazzari A, Berta L, Benedusi-Pagliano E, Frairia R (1993) Biological relevance of the interaction between sex steroid binding protein and its receptor of MCF-7 cells: effect on the estradiol-induced cell proliferation. J Steroid Biochem Mol Biol 45:435–444

Freiss G, Vignon F (1994) Antiestrogens increase protein tyrosine phosphatase activity in human breast cancer cells. Mol Endocrinol 8:1389–1396

Fujimoto N, Katzenellenbogen BS (1994) Alteration in the agonist/antagonist balance of antiestrogens by activation of protein kinase A signaling pathways in breast cancer cells: antiestrogen-selectivity and promoter-dependence. Mol Endocrinol 8:296–304

Gaub M, Bellard M, Scheuer I, Chambon P, Sassone-Corsi P (1990) Activation of the ovalbumin gene by the estrogen receptor involves the fos-jun complex. Cell 53:1267–1276

Gottardis MM, Jordan VC (1988) Development of tamoxifen-stimulated growth of MCF-7 tumors in athymic mice after long-term antiestrogen administration. Cancer Res 48:5183–5187

Groul DJ, Altschmeid J (1993) Synergistic induction of apoptosis with gluco-corticoids and 3',5'-cyclic adenosine monophosphate reveals agonist activity by RU486. Mol Endocrinol 7:104–113

Harlow KW, Smith DN, Katzenellenbogen JA, Greene GL, Katzenellenbogen BS (1989) Identification of cysteine-530 as the covalent attachment site of an affinity labeling estrogen (ketononestrol aziridine) and antiestrogen (tamoxifen aziridine) in the human estrogen receptor. J Biol Chem 264:17476–17485

Ince BA, Zhuang Y, Wrenn CK, Shapiro DJ, Katzenellenbogen BS (1993) Powerful dominant negative mutants of the human estrogen receptor. J Biol Chem 268:14026–14032

Ince BA, Schodin DJ, Shapiro DJ, Katzenellenbogen BS (1995) Repression of endogenous estrogen receptor activity in MCF-7 human breast cancer cells by dominant negative estrogen receptors. Endocrinology 136:3194–3199

Jordan VC, Murphy CS (1990) Endocrine pharmacology of antiestrogens as antitumor agents. Endocr Rev 11:578–610

Katzenellenbogen BS, Miller MA, Mullick A, Sheen YY (1985) Antiestrogen action in breast cancer cells: modulation of proliferation and protein synthesis, and interaction with estrogen receptors and additional antiestrogen binding sites. Breast Cancer Res Treat 5:231–243

Katzenellenbogen BS, Norman MJ (1990) Multihormonal regulation of the progesterone receptor in MCF-7 human breast cancer cells: interrelationships among insulin/insulin-like growth factor-I, serum, and estrogen. Endocrinology 126:891–898

Katzenellenbogen BS, Bhardwaj B, Fang H, Ince BA, Pakdel F, Reese JC, Schodin DJ, Wrenn CK (1993) Hormone binding and transcription activation by estrogen receptors: analyses using mammalian and yeast systems. J Steroid Biochem Mol Biol 47:39–48

Katzenellenbogen BS, Montano MM, Le Goff P, Schodin DJ, Kraus WL, Bhardwaj B, Fujimoto N (1995) Antiestrogens: mechanisms and actions in target cells. J Steroid Biochem Mol Biol 53: (in press)

Ke F, Ramirez VD (1987) Membrane mechanism mediates progesterone stimulatory effect on LHRH release from superfused rat hypothalami in vitro. Neuroendocrinol Endocrinol 45:514–517

Ke F, Ramirez VD (1990) Binding of progesterone to nerve cell membranes of rat brain using progesterone conjugated to ^{125}I-bovine serum albumin as a ligand. J Neurochem 54:467–472

Kim K, Ramirez VD (1986) Effects of prostaglandin E_2, forskolin and cholera toxin on cAMP production and in vitro LH-RH release from the rat hypothalamus. Brain Res 386:258–265

Kraus WL, Montano MM, Katzenellenbogen BS (1993) Cloning of the rat progesterone receptor gene 5' region and identification of two functionally distinct promoters. Mol Endocrinol 7:1603–1616

Kumar V, Green S, Staub A, Chambon P (1986) Localization of the oestradiol-binding and putative DNA binding domains of the human oestrogen receptor. EMBO J 5:2231–2236

Kumar V, Green S, Stack G, Berry M, Jin JR, Chambon P (1987) Functional domains of the human estrogen receptor. Cell 51:941–951

Lahooti H, White R, Danielian PS, Parker MG (1994) Characterization of ligand-dependent phosphorylation of the estrogen receptor. Mol Endocrinol 8:182–188

Le Goff P, Montano MM, Schodin DJ, Katzenellenbogen BS (1994) Phosphorylation of the human estrogen receptor: identification of hormone-regulated sites and examination of their influence on transcriptional activity. J Biol Chem 269:4458–4466

Lieberherr M, Grosse B, Kachkache M, Balsan S (1993) Cell signaling and estrogens in female rat osteoblasts: a possible involvement of unconventional nonnuclear receptors. J Bone Miner Res 8(11):1365–1376

Montano MM, Muller V, Katzenellenbogen BS (1994) Role for the carboxy-terminal F domain of the estrogen receptor in transcriptional activity of the receptor and in the effectiveness of antiestrogens as estrogen antagonists. Endocrinology [Suppl] 134:323 (abstract 491)

Montano MM, Mueller V, Trobaugh A, Katzenellenbogen BS (1995) The carboxy-terminal F domain of the human estrogen receptor: role in the transcriptional activity of the receptor and the effectiveness of antiestrogens as estrogen antagonists. Mol Endocrinol 9:814–825

Montminy MR, Sevarino KA, Wagner JA, Mandel G, Goodman RH (1986) Identification of a cyclic-AMP-responsive element within the rat somatostatin gene. Proc Natl Acad Sci USA 83:6682–6686

Osborne CK, Coronado E, Allred DC, Wiebe V, DeGregorio M (1991) Acquired tamoxifen resistance: correlation with reduced breast tumor levels of tamoxifen and isomerization of trans-4-hydroxytamoxifen. J Natl Cancer Inst 83:1477–1482

Pakdel F, Katzenellenbogen BS (1992) Human estrogen receptor mutants with altered estrogen and antiestrogen ligand discrimination. J Biol Chem 267:3429–3437

Pakdel F, Le Goff P, Katzenellenbogen BS (1993a) An assessment of the role of domain F and PEST sequences in estrogen receptor half-life and bioactivity. J Steroid Biochem Mol Biol 46:663–672

Pakdel F, Reese JC, Katzenellenbogen BS (1993b) Identification of charged residues in an N-terminal portion of the hormone binding domain of the human estrogen receptor important in transcriptional activity of the receptor. Mol Endocrinol 7:1408–1417

Pappas TC, Gametchu B, Watson CS (1995) Membrane estrogen receptors identified by multiple antibody labeling and impeded-ligand binding. FASEB J 9:404–410

Reese JR, Katzenellenbogen BS (1991) Differential DNA-binding abilities of estrogen receptor occupied with two classes of antiestrogens: studies using human estrogen receptor overexpressed in mammalian cells. Nucleic Acids Res 19:6595–6602

Reese JC, Katzenellenbogen BS (1992a) Characterization of a temperature-sensitive mutation in the hormone binding domain of the human estrogen receptor: studies in cell extracts and intact cells and their implications for hormone-dependent transcriptional activation. J Biol Chem 267:9868–9873

Reese JC, Katzenellenbogen BS (1992b) Examination of the DNA binding abilities of estrogen receptor in whole cells: implications for hormone-independent transactivation and the action of the pure antiestrogen ICI164,384. Mol Cell Biol 12:4531–4538

Santen R, Manni A, Harvey H, Redmond C (1990) Endocrine treatment of breast cancer in women. Endocr Rev 11:221–265

Sartorius CA, Tung L, Takimoto GS, Horwitz KB (1993) Antagonist-occupied human progesterone receptors bound to DNA are functionally switched to transcriptional agonists by cAMP. J Biol Chem 268:9262–9266

Schüle R, Rangarajan P, Kliewer S, Ransone LJ, Bolado J, Yang N, Verma IM, Evans RM (1990) Functional antagonism between oncoprotein C-jun and the glucocorticoid receptor. Cell 62:1217–1226

Sheffield LG, Welsch CW (1985) Cholera-toxin-enhanced growth of human breast cancer cell lines in vitro and in vivo: interaction with estrogen. Int J Cancer 36:479–483

Shemshedini L, Knauthe R, Sassone-Corsi P, Pornon A, Gronemeyer H (1991) Cell-specific inhibitory and stimulatory effects of fos and jun on transcription activation by nuclear receptors. EMBO J 10:3839–3849

Silberstein GB, Strickland PS, Trumpbour V, Coleman S, Daniel CW (1984) cAMP stimulates growth and morphogenesis of mouse mammary ducts. Proc Natl Acad Sci USA 81:4950–4954

Strähle U, Schmid W, Schütz G (1988) Synergistic action of the glucocorticoid receptor with transcription factors. EMBO J 7:3389–3395

Sumida C, Pasqualini JR (1989) Antiestrogen antagonizes the stimulatory effect of epidermal growth factor on the induction of progesterone receptor in fetal uterine cells in culture. Endocrinology 124:591–597

Sumida C, Pasqualini JR (1990) Stimulation of progesterone receptors by phorbol ester and cyclic AMP in fetal uterine cells in culture. Mol Cell Endocrinol 69:207–215

Sumida C, Lecerf F, Pasqualini JR (1988) Control of progesterone receptors in fetal uterine cells in culture: effects of estradiol, progestins, antiestrogens, and growth factors. Endocrinology 122:3–11

Szego CM, Davis JS (1967) Adenosine 3',5'-monophosphate in rat uterus: acute elevation by estrogen. Proc Natl Acad Sci USA 58:1711–1718

Tzukerman MT, Esty A, Santiso-Mere D, Danielian P, Parker MG, Stein RB, Pike JW, McDonnell DP (1994) Human estrogen receptor transactivational capacity is determined by both cellular and promoter context and mediated by two functionally distinct intramolecular regions. Mol Endocrinol 94:21–30

Wrenn CK, Katzenellenbogen BS (1993) Structure-function analysis of the hormone binding domain of the human estrogen receptor by region-specific mutagenesis and phenotypic screening in yeast. J Biol Chem 268:24089–24098

Yang-Yen H, Chambard J, Sun Y, Smeal T, Schmidt TJ, Drouin J, Karin M (1990) Transcriptional interference between C-jun and the glucocorticoid receptor: mutual inhibition of DNA binding due to direct protein-protein interaction. Cell 62:1205–1215

3 Analysis of Genetically Altered Mice Without Glucocorticoid Receptor

W. Schmid, T. Cole, J. Blendy, L. Montoliu, and G. Schütz

3.1   Introduction 51
3.2   Physiological Effects of Glucocorticoids 52
3.3   The Structure of the GR Gene and Its Disruption
      by Gene Targeting 54
3.4   Effects of Disruption of the GR Gene 56
3.5   Cell-Specific and Developmental Activation of a Glucocorticoid-
      Responsive Gene: The Tyrosine Aminotransferase Gene 58
3.6   Conclusion 61
References 62

3.1 Introduction

Steroid hormones regulate a number of developmental and physiological processes in vertebrates by controlling the transcriptional activity of specific genes (Beato 1989, Tsai and O'Mally 1994). The ability of target cells to respond is attributed to the presence of specific receptors which mediate the action of the hormone within the cell. The receptors are localized within the nucleus in association with other proteins, which in absence of the hormone keep the receptor in an inactive state. After binding of the hormone, the hormone–receptor complex, as a dimer, binds to specific DNA sequences. The various functions of the receptor – DNA binding, ligand binding, transcriptional activation – have been assigned to separate domains of the receptor. The unliganded receptor is maintained in a nonfunctional form by oligomerization with other proteins, among which hsp 90 is best characterized. This interaction is mediated by the ligand-binding domain. Upon binding of the hormone the receptor attains the property of specifically binding to its recognition sequence. The C-domain which is responsible for specific DNA recognition consists of two zinc fingers and is most highly conserved. Two activating functions in the amino terminal and carboxy terminal part of the protein mediate the effect of the receptor on transcription.

The steroid receptors bind to DNA by interacting with residues in the major groove of DNA where they recognize specific base pairs. Many years ago, we demonstrated that the glucocorticoid receptor (GR) and other steroid receptors bind to palindromic sequences with the characteristic sequence motif AGAACA (glucocorticoid response element, GRE; Strähle et al. 1987), whereas the estradiol receptor binds to the related motif AGGTCA (Klock et al. 1987), again, in palindromic organization. The steroid receptors bind as homodimers, whereas others, for example, the vitamin D and retinoic acid receptors, bind as heterodimers with retinoic acid X receptor (RXR) to DNA (Leid et al. 1992), (Mangelsdorf et al. 1992).

3.2 Physiological Effects of Glucocorticoids

Many different processes are controlled by glucocorticoid hormones. These include control of carbohydrate and lipid metabolism, modulation of immune responses and effects in the brain including the stress response and the feedback regulation of the hypothalamic–pituitary–adrenal (HPA) axis. Figure 1 summarizes some of the known functions of glucocorticoids. Glucocorticoids are involved in the activation of genes coding for gluconeogenic enzymes in liver and kidney. Following birth the ensuing hypoglycaemia leads to an increase in the level of glucocorticoids and glucagon, the action of which is mediated by cAMP (Pilkis und Granner 1992). The increase of these two hormones results in activation of genes encoding gluconeogenic enzymes, such as glucose-6-phosphatase, phosphoenolpyruvate carboxykinase, tyrosine aminotranferase, and serine dehydratase.

Glucocorticoids play an important role in the maturation of the lung. They increase the activation of the genes encoding surfactant proteins

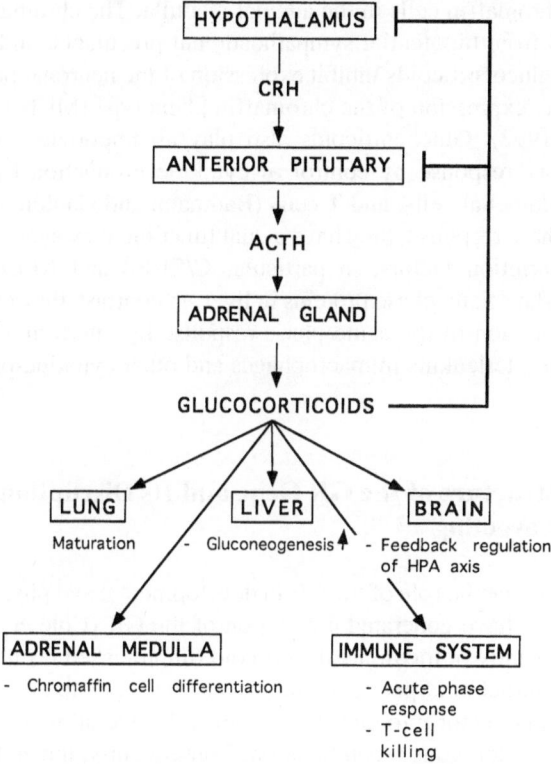

Fig. 1. Corticotropin-releasing hormone regulates secretion of adrenocortico-tropin (*ACTH*) which, in turn, stimulates the adrenal gland to release glucocorticoids. Glucocorticoids affect many different target tissues. Glucocorticoids regulate their own synthesis by negative feedback control acting at the level of hypothalamus and anterior pituitary, respectively. *CRH*, Corticotropin-releasing hormone

(Mendelson and Boggaram 1991), which reduce the surface tension of the alveolar membranes thus promoting inflation and ventilation of the lung in the perinatal period. In addition, glucocorticoids induce the transcription of the gene encoding the amiloride-sensitive sodium channel which is responsible for fluid resorption in the lung during the perinatal period (Champigny et al. 1994). Furthermore, glucocorticoids are thought to play an important role in the differentiation and determination of chromaffin cells in the adrenal medulla. The chromaffin cells are derived from bipotential sympathoadrenal precursors, and it is believed that glucocorticoids inhibit expression of the neuronal phenotype and promote expression of the chromaffin phenotype (Michelsohn und Anderson 1992). Glucocorticoids also play an important role in the inflammatory response by control of cytokine production by macrophages, endothelial cells, and T cells (Baumann and Gauldie 1994). In the acute phase response, they have a dual function: they synergize with other transcription factors, in particular C/EBPβ and NFκB, in the synthesis of the acute phase proteins in liver; in contrast, they contribute to the termination of the acute phase response by interfering with the synthesis of interleukins in macrophages and other cytokine-producing cells.

3.3 The Structure of the GR Gene and Its Disruption by Gene Targeting

In order to define the role of the GR in developmental and physiological processes, we have generated a mutation of the GR (Cole et al. 1995) and more recently of the mineralocorticoid (unpublished results) receptor by gene targeting. Specific mutations can be generated by introducing a targeting vector into embryonic stem cells and allowing the cells to undergo homologous recombination. Consequently, the endogenous gene is replaced or mutated by the introduced vector containing the neomycin resistance gene. As homologous recombination is a rare event, only a few of the cells which have inserted the targeting vector in the DNA have done so at the homologous site. Therefore, the neomycin-resistant clones has to be screened by Southern blot analysis. After identification of cells carrying a mutation at the GR locus, these embryonic stem cells were injected into the blastocyst of mice. Injected blastocysts were introduced into pseudopregnant females from which chimeras were then derived. Since the embryonic stem cells had contributed to the germ line, heterozygous and homozygous mice with a mutation at the receptor locus could easily be derived.

To generate a mutation by homologous recombination we first characterized the GR gene (Strähle et al. 1992). The gene spans over 110 kb and contains nine exons (Fig. 2) The two zinc fingers of the DNA

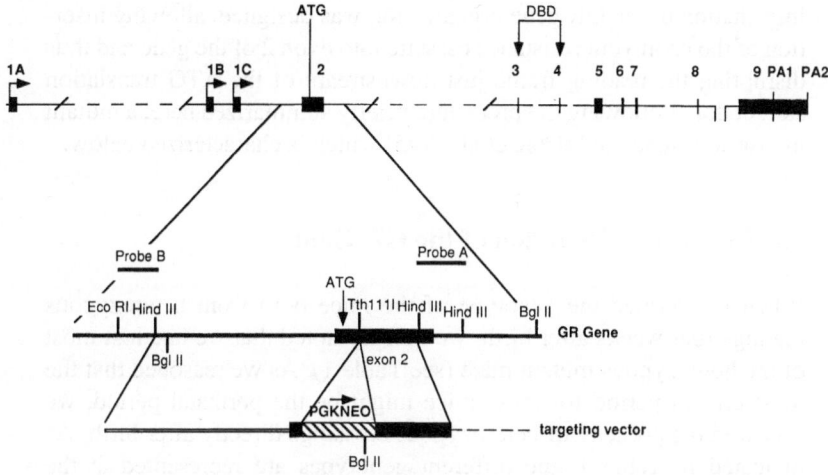

Fig. 2. Structure of the mouse glucocorticoid receptor (*GR*) gene. Genomic organization of the eight coding exons (2–9) and three alternative 5' exons (*1A-1C*) of the mouse GR gene is shown in the *uppermost panel. Horizontal arrows* indicate different start sites of transcription, *PA1* and *PA2* represent two putative polyadenylation signals. The position of the *ATG* start site of translation, exons encoding the DNA-binding domain (*DBD*), and exons encoding the hormone-binding domain (HBD) are also marked. *Below*, both the targeting construct as well as the resulting replacement are depicted. Exon 2 of the GR gene is interrupted by a neomycin resistance cassette ligated in frame with the amino terminus of the GR gene

binding domain are encoded by exons 3 and 4. The hormone-binding domain is contained in the carboxy terminal part of the receptor which is divided among the last five exons (exons 5–9). The aminoterminal part of the protein which contains a transcriptional activation domain is encoded by a single exon. Three promoters control the expression of the GR gene. One of these promoters appears to be activated only in T lymphocytes and lymphatic tissues; the other two promoters are active to somewhat different degrees in many of the cell lines and tissues which have been tested. Expression studies of the receptor during mouse development have revealed that, very surprisingly, the gene is already activated at day 10 of fetal development. For inactivation of the GR gene of the mouse a replacement targeting vector based on this information of the GR gene organization was designed, allowing insertion of the neomycin resistance cassette into exon 2 of the gene and thus disrupting the reading frame just downstream of the ATG translation start codon. Following the procedure briefly summarized here, a mutant mouse was generated (Cole et al. 1995) which is characterized below.

3.4 Effects of Disruption of the GR Gene

When we defined the genotype of the mice born from heterozygous matings four weeks after birth, we quickly noted that we had lost most of the homozygous mutant mice (see Table 1). As we reasoned that the most critical period for these mice might be the perinatal period, we followed the progeny of heterozygous crossings directly after birth. As indicated in Table 1, the different genotypes are represented at the expected mendelian frequency. We also noticed that we lost most of them shortly after birth. A possible cause of death was indicated by the appearance of the homozygous mutant mice: they were cyanotic, showed forced breathing, and died within the first hours after birth. The mice displaying acute respiratory distress were histologically analyzed. Comparison of the lungs of wild-type with GR-deficient mice indicated severe lung atelectasis in the mutant animals. As more recent studies have shown, this might, at least in part, be due to reduced synthesis of surfactant protein C (Mendelson and Boggaram 1991) and to reduced expression of the amiloride-sensitive sodium channel which is responsible for absorption of alveolar fluid around birth (Champigny et al. 1994).

As already alluded to, glucocorticoids are required for gluconeogenesis. Comparison of the messenger RNA levels coding for key glucone-

Table 1. Genotypes of progeny of heterozygote intercrosses

Progeny	Genotypes	+/+	+/–	–/–
Nonisogenic	4 weeks	188 (35%)	331 (61%)	25 (4.5%)
	Born	26 (21%)	69 (56%)	28 (23%)
	Alive after 4 h	26 (21%)	69 (56%)	2 (1.6%)
Isogenic	4 weeks	30	48	0
129 sv	d18.5	6	8	7

ogenic enzymes revealed that newborn livers from homozygous mice had reduced levels for key gluconeogenic enzymes, such as glucose-6-phosphatase and phosphoenolpyruvate carboxykinase.

The synthesis of glucocorticoids in the zona fasciculata of the adrenal cortex is under the control of the pituitary. The level of glucocorticoids is regulated by a feedback control mechanism which operates at the level of the pituitary as well as of the hypothalamus which controls the synthesis of adrenocorticotropin (ACTH) in anterior pituitary by secretion of corticotrophin releasing hormone (CRH; Chrousos and Gold 1992). We noticed that the adrenal glands from homozygous mutant mice are about twice the size of those of wild-type littermates due to hypertrophy and hyperplasia of adrenal cortical cells. This hyperplasia is due to strongly increased levels of ACTH, which were found to be 15- to 20-fold higher in the mutants. The increased levels of ACTH led to a threefold increase in the serum levels of corticosterone. The hypertrophy of the adrenal cortex as a consequence of increased stimulation via the defective negative feedback resulted in increased expression of key cortical steroid biosynthetic enzymes, e.g., $P\ 450_{scc}$, $P450_{c11\beta}$.

As indicated above, glucocorticoids are thought to play an important role in the differentiation of neural crest-derived chromaffin cells from bipotential sympathoadrenal progenitors (Michelsohn and Anderson 1992; Anderson 1993). The major function of chromaffin cells is the synthesis and secretion of epinephrine and norepinephrine. Is differentiation and proliferation/survival of chromaffin cells affected by the receptor mutation? Histological analysis and immunohistochemical studies using antibodies against synaptophysin and tyrosine hydroxylase, which are markers for chromaffin cells, and against the epinephrine-producing enzyme phenylethanolamine N-methyltransferase (PNMT) revealed that the adrenal medulla is abnormal. We could detect only a small number of scattered chromaffin cells. No PNMT-positive cells were identified, which suggested absence of epinephrine-producing cells. This enzyme which catalyzes the conversion of norepinephrine to epinephrine is known to be regulated by glucocorticoids (Ross et al. 1990). Indeed, no epinephrine could be found in mutant adrenals by high performance liquid chromatography (HPLC) or hormone determination by radioimmunoassay (RIA).

These results show the presence of only norepinephrine-producing and the complete absence of epinephrine-producing chromaffin cells in the adrenal gland of homozygous mutant mice and define a crucial role of glucocorticoid in the maturation of the adrenal medulla. Furthermore, these data suggest the possible existence of a separate precursor for epinephrine-producing chromaffin cells. Noradrenergic chromaffin cells are apparently not strictly dependent on the glucocorticoid signaling pathway.

3.5 Cell-Specific and Developmental Activation of a Glucocorticoid-Responsive Gene: The Tyrosine Aminotransferase Gene

Transfection studies as well as analysis of protein–DNA interactions have allowed the control elements of the tyrosine amino transferase (TAT) gene to be defined (Jantzen et al. 1987; Ruppert et al. 1990; Nitsch et al. 1990; Weih et al. 1990). Three enhancers have been found to be required for high-level liver cell-specific expression of the gene in dependence of external signals (Fig. 3) but each of the enhancers is able to function as a separate unit at a lower level of activity. A set of transcription factors which contribute to liver-specific expression as well as to hormonal regulation of the gene has been identified in the two proximal enhancers (Nitsch et al. 1993). The glucocorticoid-inducible enhancer, 2.5 kb upstream of the start site of transcription, contains a GRE besides binding sites for factors determining liver cell-specific activity of the enhancer (HNF3 and C/EBP). The enhancer at 3.6 kb pairs upstream of the start site contains a cAMP-responsive element (CRE), besides binding sites for liver-enriched factors, in particular HNF4, a member of the steroid receptor supergene family. The enhancer 11 kb upstream of the promotor further contributes to liver cell-specific activity. The activity of the hormone response elements and of the regulatory elements determining liver cell-specific expression are strictly interdependent. This modular organization guarantees responsiveness to ubiquitous hormonal signals exclusively in specific cell types, here, the liver parenchymal cells.

The principle of modular composition of cell-specific and hormone-controlled enhancers, first established in cells in culture, has now been

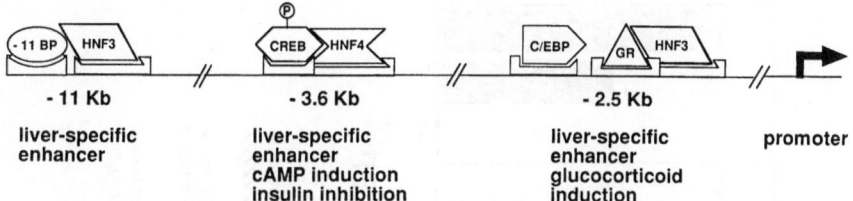

Fig. 3. Upstream enhancers controlling the expression of the tyrosine amino-transferase gene. The three enhancers are situated 5' of the promotor at −2.5, −3.6, and −11 kb, respectively. The *symbols* indicate binding of specific transcription factors. These are hepatocyte nuclear factor 3 (*HNF-3*), the glucocorticoid receptor (*GR*), the CAAT/enhancer-binding protein (*C/EBP*), hepatocyte nuclear factor 4 (*HNF-4*), the cAMP response element-binding protein (*CREB*), and a −11 site-specific-binding protein (*−11BP*)

confirmed by in vivo experiments in transgenic mice. The use of signal-dependent constructs in transgenic mice has clarified the role of the hormones in the hormonal and developmental expression of the TAT gene (Montoliu et al. 1995). These studies also suggest that the coordinated control by ubiquitous signal transduction pathways might play a major role in the developmental control of the TAT and other gluconeogenetic genes. Glucocorticoids and glucagon, the action of which is mediated by cAMP, are crucial for the activation of the gluconeogenic pathway after birth. Due to the hypoglycemia, which is the result of interrupted maternal blood supply, genes encoding gluconeogenic enzymes are activated. Hypoglycemia leads to a rise in the levels of glucocorticoids and glucagon which enhance gluconeogenesis, whereas insulin levels which have an inhibitory role are downregulated. As a consequence the transcription of genes involved in gluconeogenesis, e.g., the TAT gene, is strongly increased.

To prove this hypothesis, we designed particular constructs, the design of which is shown in the lower part of Fig. 4. In front of the TATA box of the thymidine kinase gene dimerized binding sites for the GR, for the cAMP response element binding protein, CREB, or for the ubiquitous transcription factor SP1 were inserted. The constructs also contained the enhancer of the α-fetal protein gene which restricts expression of the transgene to liver parenchymal cells. As expected, the construct containing binding sites for the constitutively active transcrip-

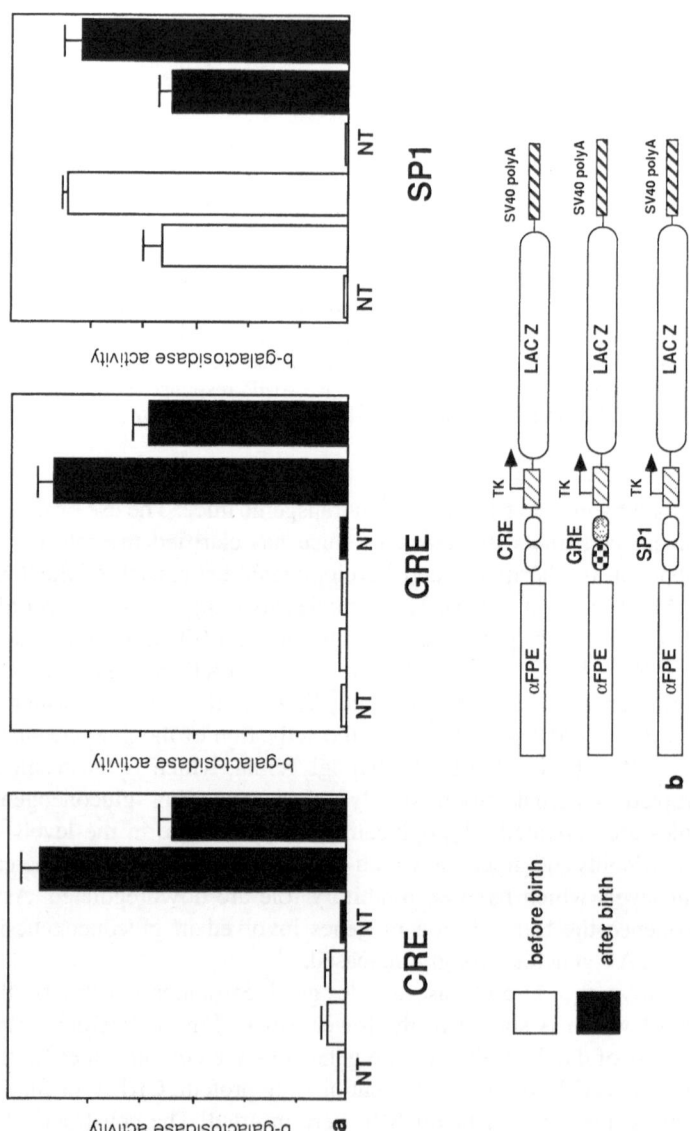

Fig. 4a,b. Legend see p. 61

tion factor SP1 was active in mouse liver before and after birth. In contrast, the CRE and GRE containing transgenes are inactive before birth, but undergo perinatal activation within hours after birth.

These results confirm the central role of glucocorticoid and CRE in developmentally controlled activation of TAT and, most likely, of other genes coding for gluconeogenetic enzymes which all have been shown to contain GREs and CREs.

3.6 Conclusion

Disruption of the GR gene by gene targeting allowed the role of this receptor to be defined physiologically and during development. Inactivation of the gene blocks adrenergic chromaffin cell development, retards lung maturation, impairs activation of expression of genes coding for gluconeogenic enzymes, and disrupts feedback regulation via the HPA axis. Cell-specific ablation of the GR in future experiments will certainly help to define the role of glucocorticoids in various physiological and pathological conditions more precisely. Furthermore, experiments in progress aim to develop mutations of the receptor which will selectively abolish activating and inhibitory functions. Mice carrying selective GR mutations will be extremely useful for understanding receptor function in the normal and diseased organism. They will also form the basis for the search of new drugs with selective effects, which interfere with the inhibitory activities of the receptor without affecting the activating functions and vice versa.

◀ **Fig. 4a,b.** Perinatal activation of chimeric *lacZ* constructs in chimeric mice. **a** Bar diagrams of two transgenic lines for each construct documents β-galactosidase activities of liver protein extracts from day 18.5 of gestation (*open bars*) and day 0.5 newborns (*solid bars*). *NT*, nontransgenic littermates before and after birth. **b** The three constructs used for the generation of the transgenic lines. *CRE*, cAMP response element; *GRE*, glucocorticoid response element; *SP1*, transcription factor

References

Anderson D (1993) Molecular control of cell fate in the neural crest. Annu Rev Neurosci 16:129–58

Baumann H, Gauldie J (1994) The acute phase response. Immunol Today 15:74–80

Beato M (1989) Gene regulation by steroid hormones. Cell 56:335–344

Champigny G, Voilley N, Lingueglia E, Friend V, Barbry P, Lazdunski M (1994) Regulation of expression of the lung amiloride-sensitive Na$^+$ channel by steroid hormones. EMBO J 13:2177–2181

Chrousos GP, Gold PW (1992) The concepts of stress and stress system disorders. JAMA 267:1244–1252

Cole T, Blendy JA, Monaghan AP, Krieglstein K, Schmid W, Aguzzi A, Fantuzzi G, Hummler E, Unsicker K, Schütz G (1995) Targeted disruption of the glucocorticoid receptor gene blocks adrenergic chromaffin cell development and severely retards lung maturation. Genes Dev (in press)

Jantzen HM, Strähle U, Gloss B, Stewart F, Schmid W, Boshart M, Miksiced R, Schütz G (1987) Cooperativity of glucocorticoid response elements located far upstream of the tyrosine aminotransferase gene. Cell 49:29–38

Klock G, Strähle U, Schütz G (1987) Estrogen and glucocorticoid responsive elements are closely related, but distinct. Nature 329:734–736

Leid M, Kaestner P, Lyons R, Nakshatri H, Saunders M, Zacharewski T, Chen J-Y, Staub A, Garnier J-M, Mader S, Chambon P (1992) Purification, cloning, and RXR identity of the HeLa cell factor with which RAR or TR heterodimerizes to bind target sequences efficiently. Cell 68:377–395

Mangelsdorf D, Borgmeyer U, Heyman R, Yang Zhou J, Ong E, Oro A, Kakizuka A, Evans R (1992) Characterization of three RXR genes that mediate the action of 9-cis retinoic acid. Genes Dev 6:329–344

Mendelson CR, Boggaram V (1991) Hormonal control of the surfactant system in fetal lung. Annu Rev Physiol 53:415–440

Michelsohn AM, Anderson DJ (1992) Changes in competence determine the timing of two sequential glucocorticoid effects on sympathoadrenal progenitors. Neuron 8:589–604

Montoliu L, Blendy JA, Cole T, Schütz G (1995) Analysis of perinatal gene expression: hormone, response elements mediate activation of a LacZ reporter gene in liver of transgenic mice. Proc Natl Acad Sci USA 92:4244–4248

Nitsch D, Stewart AF, Boshart M, Mestril R, Weih F, Schütz G (1990) Chromatin structures of the rat tyrosine aminotransferase gene relate to the function of its cis-acting elements. Mol Cell Biol 10:3334–3342

Nitsch D, Boshart M, Schütz G (1993) Activation of the tyrosine aminotransferase gene is dependent on synergy between liver-specific and hormone-responsive elements. Proc Natl Acad Sci USA 90:5479–5483

Pilkis SJ, Granner DK (1992) Molecular physiology of the regulation of hepatic gluconeogenesis and glycolysis. Annu Rev Physiol 54:885–909

Ross ME, Evinger MJ, Hyman SE, Carroll JM, Mucke L, Comb M, Reis DJ, Joh TH, Goodman HM (1990) Identification of a functional glucocorticoid response element in the phenylethanolamine N-methyltransferase promoter using fusion genes introduced into chromaffin cells in primary culture. J Neurosci 10:520–530

Ruppert S, Boshart M, Bosch FX, Schmid W, Fournier REK, Schütz G (1990) Two genetically defined trans-acting loci coordinately regulate overlapping sets of liver-specific genes. Cell 61:895–904

Strähle U, Klock G, Schütz G (1987) A DNA sequence of 15 base pairs is sufficient to mediate both glucocorticoid and progesterone induction. Proc Natl Acad Sci USA 84:7871–7875

Strähle U, Schmid A, Kelsey G, Stewart AF, Cole TJ, Schmid W, Schütz G (1992) At least three promoters direct expression of the mouse glucocorticoid receptor gene. Proc Natl Acad Sci USA 89:6731–6735

Tsai M-J, O'Malley BW (1994) Molecular mechanisms of action of steroid/thyroid receptor superfamily members. Annu Rev Biochem 63:451–486

Weih F, Stewart AF, Boshart M, Nitsch D, Schütz G (1990) In vivo monitoring of cyclic AMP-stimulated DNA binding activity. Genes Dev 4:1437–1449

4 Organ-Selective Actions of Tamoxifen and Other Partial Antiestrogens

R. T. Turner

4.1	Introduction	65
4.2	Gonadal Function and Bone Mass	66
4.2.1	Bone as an Estrogen Target Tissue	66
4.2.2	Estrogen and Bone Cell Function	68
4.2.3	Role of Estrogen Receptors	70
4.2.4	Mechanism of Estrogen Action on the Skeleton	71
4.3	Estrogen Analogs	72
4.3.1	Antiestrogens and Bone Mass	72
4.3.2	Mechanisms of Action of Tamoxifen	77
4.3.3	Bone Cell-Selective Actions of Estrogen Analogs	78
4.4	Summary	79
References		79

4.1 Introduction

A serious obstacle to the rational development of innovative approaches for preventing and/or treating osteoporosis is the idiopathic nature of postmenopausal bone loss. Estrogen deficiency is an important risk factor for osteoporosis. However, not all postmenopausal women develop osteoporotic fractures, indicating that estrogen deficiency alone is insufficient to fully account for the disorder. An increased understanding of the mechanisms whereby estrogen interacts with other factors to maintain bone balance is an important step in improving hormone replacement therapy in postmenopausal women.

Estrogen analogs with different degrees of agonist and antagonist activities were initially used as tools for investigating the mechanism of estrogen action on target tissues, including the skeleton. However, the results of the early studies suggested that these agents might offer important advantages over natural estrogens for hormone replacement therapy to prevent or treat postmenopausal osteoporosis.

For the purpose of this manuscript, mixed estrogen agonist/antagonists represent synthetic analogs of estrogen which antagonize the actions of endogenous and/or administered natural estrogens on specific estrogen-mediated actions. In the case of tamoxifen this antagonism has proven to be clinically useful to arrest the growth of breast tumors. This class of compounds typically has some estrogenic activity, especially when circulating endogenous estrogen levels are low. Of clinical interest is the question of whether the respective levels of agonism and antagonism vary between organs The evidence which supports organ-selective actions of estrogen agonists is reviewed in this chapter, emphasizing the potential value of these agents for prevention and treatment of postmenopausal osteoporosis.

4.2 Gonadal Function and Bone Mass

4.2.1 Bone as an Estrogen Target Tissue

Bone is a target of estrogen action. The hormone is important for normal (a) sexual dimorphism of the skeleton, (b) epiphyseal closure, (c) establishment of peak bone mass, (d) maintenance of mineral homeostasis during pregnancy and lactation in placental mammals and during the egg laying cycle in oviparous vertebrates, and (e) maintenance of bone balance in adults (reviewed by Turner et al. 1994b). The latter function of the hormone has been emphasized in interpreting the role of estrogen in the pathogenesis of postmenopausal osteoporosis. Estrogen deficiency leads to an increase in bone turnover and an imbalance between bone formation and bone resorption such that the latter predominates and, as a consequence, bone mass is reduced. As bone mass declines, the risk for bone fractures increases (Riggs et al. 1981).

It is becoming increasingly apparent that although postmenopausal osteoporosis is a serious clinical problem (Turner et al. 1994b; Cummings et al. 1985; Garn et al. 1967; Riggs et al. 1986; Arnold 1973; Aitken et al. 1973), it is best approached within the context of the multiorgan changes occurring at menopause. As an example, coronary heart disease is infrequent in premenopausal women, but the rate rises, due in part to the lower serum estrogen levels, following menopause to become the leading cause of death in women, far exceeding the mortality and morbidity from breast cancer, uterine cancer, and bone fractures combined (Mack and Ross 1989).

Long-term estrogen replacement is effective in reducing bone turnover, bone loss, and fracture risk. Observational data strongly suggest that long-term estrogen therapy reduces coronary heart disease as well as postmenopausal bone loss (Aitken et al. 1973; Mack and Ross 1989; Hazzard 1989; Genant et al. 1989). Other proposed approaches to treat postmenopausal osteoporosis include inhibitors of bone turnover (e.g., bisphosphonates, gallium nitrate, and calcitonin) and specific activators of bone formation (e.g., anabolic steroids and parathyroid hormone). Although these agents may become important in treatment of osteoporosis, they do not address the problem of an increased risk for cardiovascular disease.

Arguably, replacement of estrogen in some form is the most desirable treatment approach for preventing postmenopausal bone loss within the context of total postmenopausal biology. There are, however, three important but interrelated concerns which limit the use and/or effectiveness of estrogen replacement therapy. These concerns include: (1) timing of replacement, (2) duration of replacement therapy, and (3) undesirable side effects.

The timing of initiation of estrogen replacement as well as duration of treatment is important. Estrogen replacement therapy to prevent postmenopausal bone loss is most effective when initiated within 3 years after cessation of menses. Later administration prevents additional bone loss due to estrogen deficiency but is largely ineffective in restoring bone that has been previously lost (Turner et al. 1994b; Cummings et al. 1985; Riggs et al. 1986; Aitken et al. 1973; Mack and Ross 1989; Genant et al. 1989). Cessation of treatment leads to increased bone loss to values comparable to those observed after menopause and may result in a loss of protection against fractures (Cummings et al. 1985; Aitken et al. 1973; Mack and Ross 1989). For maximal protection against osteoporosis, estrogen should be given to asymptomatic women and treatment may require maintenance for 20 or more years!

The dual requirements for early administration of estrogen and long duration of treatment with the hormone are important concerns because of the numerous side effects of estrogen replacement therapy, which include, but are not limited to, breast soreness, breakthrough bleeding, cystic mastitis, endometrial hyperplasia, endometrial cancer, gallstones, return of menses, weight gain, and breast cancer (Turner et al. 1994b; Mack and Ross 1989; Henderson 1989). The significant possibility of life-threatening side effects is especially worrisome because many post-menopausal women electing treatment would not have, if untreated, developed bone fractures.

Synthetic estrogen analogs may prove to be greatly superior to estrogen if they are either more efficacious than the natural hormone or if fewer undesirable side effects occur with treatment. It is interesting to note that many of the most important undesirable side effects of estrogen are due to hormonal stimulation of reproductive tissues. As a consequence, analogs which selectively target the skeleton and liver but not reproductive tissues are desirable. Furthermore, analogs which improve the balance between bone formation and bone resorption would be preferred over the marked inhibition of both processes conferred by estrogen. Clinical studies of tamoxifen and raloxifene and work with these and other estrogen analogs in laboratory animal models provide compelling evidence that mixed estrogen agonists/antagonists have tissue-selective actions. This finding suggests that tissue-selective estrogen agonists offer an exciting new approach to estrogen replacement therapy.

4.2.2 Estrogen and Bone Cell Function

The relationship between bone resorption by osteoclasts and bone formation by osteoblasts is the starting point for achieving and maintaining bone balance. During growth, bone balance is positive, being overwhelmingly in the direction of bone formation. In young adults these opposing processes are nearly equal (zero bone balance), and in older individuals bone resorption exceeds bone formation (negative bone balance). Menopause in humans (Cummings et al. 1985; Garn et al.

1967; Riggs et al. 1986; Aitken et al. 1973; Eastell et al. 1988) and ovariectomy in both humans (Cummings et al. 1985; Aitken et al. 1973) and rats (Wronski et al. 1985, 1986; Turner et al. 1987a) results in increases in bone resorption and to a lesser extent bone formation, leading to an overall loss of endocortical and cancellous bone (Wronski et al. 1985, 1986; Turner et al. 1987a, 1993b; Kalu et al. 1989). Most studies suggest that estrogen treatment prevents osteopenia in adult humans (Turner et al. 1994b; Aitken et al. 1973; Genant et al. 1989) and rats (Turner et al. 1987b, 1993b) by decreasing bone turnover. Although the net overall effect of estrogen to stabilize cancellous bone volume is well established, the precise step-by-step process whereby the hormone modifies the bone remodeling cycle has been poorly characterized in humans and in laboratory animals.

Cancellous bone remodeling occurs when a foci of resorption (remodeling unit) is initiated on a previously quiescent trabecular surface (Frost 1963). A unit of bone is removed and the resulting lacuna is filled shortly thereafter by new bone formation. The formation phase of the bone remodeling sequence can theoretically underfill, precisely fill, or overfill the lacuna, resulting in a small decrease, no change, or small increase of the trabecular volume, respectively. The overall rate of bone remodeling is primarily determined by the number of remodeling foci. During the bone remodeling sequence, bone formation follows bone resorption and as a consequence is coupled to bone resorption. Thus, because of coupling, changes in resorption are followed by corresponding increases or decreases in bone formation.

Another type of bone turnover allows bone formation and bone resorption to be independently changed in magnitude on cancellous and endocortical bone surfaces, leading to architectural changes. This process is called modeling and need not involve a coupled response between bone formation and resorption (Frost 1963). The individual contributions of altered bone modeling and remodeling to the inhibition of bone turnover by estrogen are not known because most studies have not attempted to distinguish between the two processes. However, the architectural changes (increase in medullary area and decrease in trabecular number with little change in trabecular thickness) which follow ovariectomy imply an important role for altered bone modeling in the etiology of estrogen deficiency bone loss (Turner et al. 1987a; Kalu et al. 1989; Miller and Wronski 1993).

The effects of estrogen in inhibiting bone resorption in the ovariectomized rat model are universally accepted (Turner et al. 1987a, 1993b;
Wronski et al. 1988; Yamamoto and Rodan 1990). In contrast, the
effects of the hormone on bone formation are controversial. Whereas
most investigators have reported that estrogen inhibits bone formation
(Turner et al. 1993b; Wronski et al. 1988), there have also been reports
that the hormone stimulates formation (Yamamoto and Rodan 1990;
Somjen et al. 1989; Chow et al. 1992). Extensive studies in the author's
laboratory failed to reveal any evidence for an anabolic action of estrogen on bone formation and consistently demonstrated a potent inhibitory action of the hormone on bone turnover (Turner et al. 1993b;
Westerlind et al. 1993).

4.2.3 Role of Estrogen Receptors

Despite the profound effects of sex steroids on bone, early studies not
only failed to find estrogen receptors in rat (Chen and Feldman 1978)
and human (Yoshioka et al. 1980) skeletal tissues, but they also reported
that bone cell and organ cultures were insensitive to estrogen (Caputo et
al. 1976). These reports led to the widely held concepts that bone cells
lacked estrogen receptor and that estrogen affected bone cells indirectly.
More recent work has refocused attention on possible direct effects of
estrogen on osteoblasts. Several laboratories have demonstrated the
presence of estrogen receptor in cells derived from skeletal tissues
(Eriksen et al. 1988; Komm et al. 1988) and primary cultures of rat bone
marrow. Estrogen receptor mRNA was identified in normal human
osteoblast-like cell strains (Eriksen et al. 1988), whereas the author's
laboratory has demonstrated estrogen receptor messenger RNA in periosteal cells from rat long bones and calvaria and has demonstrated
estrogen binding to endocortical lining cells known to respond to estrogen in quail (Turner et al. 1993a).

Extrapolation from studies of other sex steroid receptors and other
target tissues suggests the likelihood that estrogen receptor number is
regulated in bone cells and such regulation influences the skeletal effects of estrogen. This potentially important line of research has been
hampered by difficulties in identifying and quantifying estrogen receptor in skeletal tissues. However, estrogen was shown to upregulate

progesterone receptors in cultured normal human osteoblast-like cells (Eriksen et al. 1988) and 1,25-dihydroxyvitamin D3 upregulated estrogen receptor expression in cultured marrow cells (Bellido et al. 1993). Interestingly, a recent study in women with breast cancer suggests that estrogen receptor number in bone increases with age (Frenay et al. 1991). Thus, although the presence of estrogen receptor in skeletal tissues seems well established, the possible role of receptor modulation in mediating physiological and pathological regulation of bone metabolism in response to steroid hormones is unknown.

4.2.4 Mechanism of Estrogen Action on the Skeleton

Cell culture experiments as well as laboratory animal studies suggest a multistep mechanism of estrogen action on target tissues, including the skeleton. A recently reviewed (Turner et al. 1994b) hypothesis is that estrogen acts on target cells primarily via an estrogen receptor-mediated cascade. In this model, the cascade response, while initially similar, later becomes unique to each type of target cell. The cascade is initiated by rapid modulation of a relatively small number of regulatory genes (early genes), the pattern of which is target cell specific. These early genes code for transcription factors and other intracellular effectors which in turn regulate expression of growth and differentiation factors (intermediate genes). These intermediate genes influence the subsequent expression of a very large number of genes (late genes) that mediate most of the observed phenotypic changes in target cell differentiation and synthetic activity.

Estrogen deficiency invariably leads to increased bone turnover but bone loss is site specific (Turner et al. 1994b). There is evidence that estrogen deficiency reduces the sensitivity of bone cells to mechanical strain such that regions of the skeleton under low strain are at greater risk to undergo net bone resorption (Westerlind et al. 1995).

Estrogen deficiency increases bone turnover and negatively influences the balance between formation and resorption of cancellous bone, resulting in rapid bone loss. The precise cellular mechanism in humans is unknown. However, in rats most of the bone loss occurs when osteoclasts destroy trabecular plates effectively "short circuiting" the bone formation phase of the bone remodeling cycle. Treatment with estrogen inhibits trabecular plate destruction in rats and also inhibits bone formation in rats and humans and thus is ineffective in either species in restoring bone that has already been lost. Furthermore, because estrogen replacement has undesirable side effects in some women, the benefits of preventative treatment with the hormone can be outweighed by the risks.

4.3 Estrogen Analogs

4.3.1 Antiestrogens and Bone Mass

Estrogen analogs have been identified (Table 1) which can be classified as having agonist, antagonist, or mixed agonist/antagonist activities. Diethylstilbestrol is an example of a nonsteroidal estrogen analog which binds to the estrogen receptor with high affinity and confers a complete estrogenic response on all estrogen target tissues investigated (Turner et al. 1993b). Tamoxifen, clomiphene, raloxifene, and droloxifene are mixed estrogen agonists/antagonists which show variable levels of estrogenic activity and tissue selection (Moon et al. 1991; Evans et al. 1994; Beall et al. 1984). ICI 182,780 (Gallagher et al. 1993) and ZM 189–154 (Dukes et al. 1994) are examples of estrogen antagonists which show little or no agonistic activities. Table 1 refers to rat data only. There can be considerable species differences. For example, tamoxifen which is a mixed estrogen agonist/antagonist in rats and hu-

Table 1. Tissue selective activities of estrogen analogs in growing rats

Analog	Uterine Growth	Bone Volume
Diethylstilbestrol	Agonist	Agonist
Tamoxifen	Antagonist	Agonist
4-Hydroxy Tamoxifen	Antagonist	Agonist
3-Hydroxy Tamoxifen (Droloxifene)	Antagonist	Agonist
Clomiphene	Antagonist	Agonist
Raloxifene	Antagonist	Agonist
ICI 182,780	Antagonist	Antagonist
Zm 189–154	Antagonist	Antagonist

mans is a pure estrogen antagonist in birds, but is primarily an estrogen agonist in mice (Jordan and Murphy 1990).

Tamoxifen is a nonsteroidal estrogen analog which by virtue of its ability to inhibit growth of mammary tumors and inhibit uterine growth in ovary-intact and estrogen-treated ovariectomized rats is classified an "antiestrogen" (Jordan and Murphy 1990). Most (Turner et al. 1987b, 1988; Jordan et al. 1987b; Kalu et al. 1991; Williams et al. 1991) but not all (Feldman et al. 1989) studies in rats have concluded that tamoxifen acts as a partial estrogen agonist on the skeleton. Tamoxifen has an intrinsic estrogen effect on certain organs (bone and liver) under conditions in which the estrogen analog has primarily antagonistic effects on others (uterus and breast).

In dose–response studies in intact and ovariectomized rats, tamoxifen is a partial estrogen agonist on bone at concentrations that result in uterine atrophy (Figs. 1–3). Tamoxifen prevented the expected increases in bone formation rate, osteoclast number, and osteoclast size as well as the decrease in cancellous bone volume in ovariectomized rats (Turner et al. 1987b, 1988). Tamoxifen also prevented the increase in osteoclast number in an immobilized limb and decreased osteoclast number in normal male rats (Wakley et al. 1988). Similarly, tamoxifen reduced cancellous osteopenia in the immobilized limb of dogs (Waters et al. 1991) and also osteoclast number and bone resorption in organ culture, suggesting that it may have a direct effect on existing osteoclasts (Stewart and Stern 1986); however our studies in rats suggest that the principal effect of either estrogen or tamoxifen is to inhibit osteoclast fusion (Turner et al. 1988, 1994a; Moon et al. 1991; Wakley et al. 1988). In contrast, the "pure" estrogen antagonist ICI 182,780 was reported to reduce cancellous bone volume in ovary-intact rats (Gallagher et al. 1993). Similar findings were obtained with ZM 189–154 (Dukes et al. 1994).

Tamoxifen antagonizes growth of estrogen-dependent breast tumor cells, is approved in the United States as an adjuvant to surgical mastectomy, and is being investigated to determine whether the drug prevents breast cancer. Tamoxifen binds to estrogen receptors and blocks the effects of endogenous estrogen in some tissues (Santen et al. 1990; Jordan and Murphy 1990) but other nonestrogen receptor-mediated effects of tamoxifen have been described (Jordan and Murphy 1990). Thus, tamoxifen is a tumorstatic agent in tumors which have estrogen

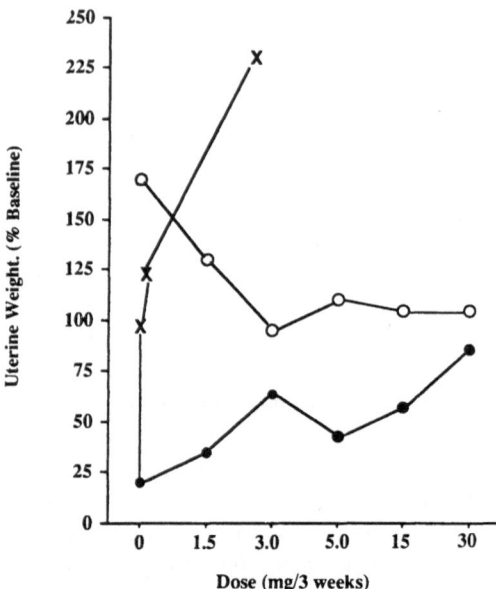

Fig. 1. The effects of ovariectomy, tamoxifen, and estrogen on uterine growth in 2-month-old Sprague-Dawley rats. The data are expressed as the percent change in weight compared to baseline control rats sacrificed on the day the treatments were started (100%). The dose indicates the total amount of tamoxifen administered to intact (*open circles*) and ovariectomized (*closed circles*) rats, and estrogen (*x*) to ovariectomized rats during the 3-week experiment. The drugs were given by subcutaneous implantation of controlled release pellets (Turner et al. 1987a) designed to release the indicated mass during the experiment. There was an increase in uterine weight compared to the baseline controls (100%) in the intact rats, whereas ovariectomy resulted in a decrease in weight (d%). Tamoxifen treatment prevented the increase in uterine weight in intact rats in a dose-dependent manner. Tamoxifene prevented the decrease in weight in the ovariectomized rats but did not increase weight compared to the initial values. Thus, tamoxifen did not promote uterine growth in ovariectomized rats and antagonized growth in ovary-intact animals. In contrast, estrogen treatment of ovariectomized rats stimulated uterine growth and high doses of the hormone (pellets, 0.1 mg) resulted in growth in excess of that for intact animals. These findings are interpreted to indicate that high doses of tamoxifen antagonize the growth promoting action of endogenous estrogens on the uterus and that the antiestrogen has only very weak estrogen agonistic effects on the uterus to prevent further uterine atrophy in ovariectomized animals. Please see Moon et al. (1991) for additional details

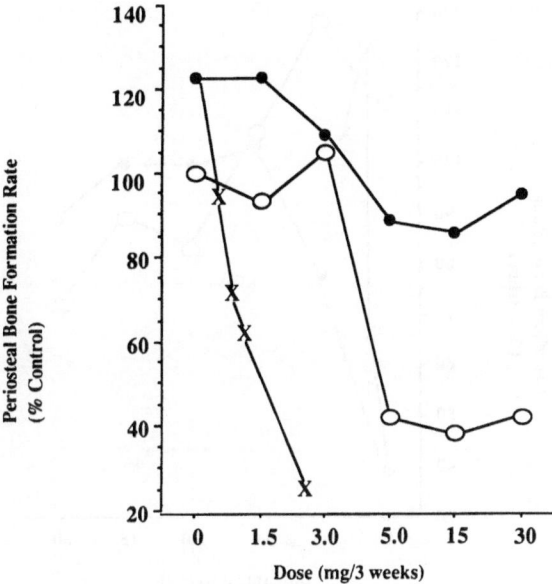

Fig. 2. The effects of ovariectomy, tamoxifen, and estrogen on the periosteal bone formation rate at the tibial diaphysis in growing rats. Tamoxifen was administered to intact (*open circles*) and ovariectomized (*closed circles*) rats, and estrogen to ovariectomized rats as described in Fig. 1. The sex difference in bone mass in rats is due, in part, to the inhibitory effects of estrogen on radial growth (Turner et al. 1989, 1990; Wakley et al. 1991). Thus, the increase in bone formation in ovariectomized rats compared to intact controls (100%) was expected. Tamoxifen treatment resulted in a dose-dependent decrease in the bone formation rate to values that did not differ significantly from the intact controls. Tamoxifen also reduced the bone formation rate in intact animals. Low doses (0.05 mg pellet) of estrogen reduced the bone formation rate in ovariectomized rats to values similar to those in intact animals but higher doses of the hormone almost completely suppressed bone formation. We interpret these results to indicate that tamoxifen acts as an estrogen agonist on bone formation but the potency and efficacy of this drug is less than that of 17β-estradiol. Furthermore, tamoxifen did not antagonize endogenous estrogens in intact rats. Antagonism would have resulted in an increase in bone formation. Instead, bone formation was further decreased in ovary-intact rats, suggesting that the effects of tamoxifen and estrogen were addictive

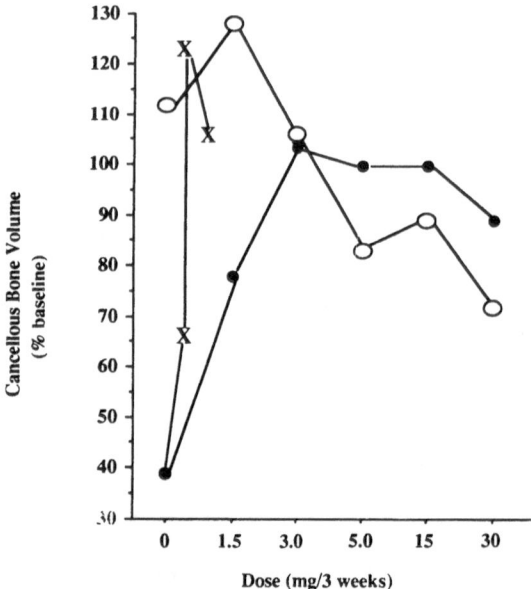

Fig. 3. The effects of ovariectomy, tamoxifen and estrogen on cancellous bone volume in 2-month-old Sprague-Dawley rats. The data are expressed as the percent change in bone volume compared to baseline control rats sacrificed on the day treatment were started. Tamoxifen was administered to intact (*open circles*) and ovariectomized (*closed circles*) rats and estrogen to ovariectomized rats as described in Fig. 1. Cancellous bone volume increased slightly in intact rats compared to the baseline (100%) animals while there was a dramatic decrease in bone volume in the ovariectomized rats. The observed cancellous bone osteopenia is a well established response to estrogen deficiency in the rat model (reviewed in Turner et al. 1994b). Tamoxifen treatment of ovariectomized rats prevented cancellous bone osteopenia. A low dose of tamoxifen resulted in an increase in cancellous bone volume compared to the baseline controls but higher doses of the drug resulted in values that did not differ from ovary-intact animals or estrogen-treated ovariectomized rats. Estrogen treatment prevented osteopenia in ovariectomized rats and the 0.1 mg pellet appeared to increase bone volume compared to the baseline control. Low doses of tamoxifen increased cancellous bone volume in ovary-intact rats to values similar to the peak value induced by estrogen. Very high doses of tamoxifen resulted in a small reduction in cancellous bone volume in ovary-intact rats. Cancellous bone volume is determined in growing rats by the balance between endochondral ossification, bone formation, and bone resorption. Since each of the processes are influenced by gonadal hormones, we cannot ascribe a specific cellular mechanism to these findings. *Cont. p. 77*

receptor and respond to estrogen. To achieve optimal prevention of tumor recurrence, patients must be treated with the drug for 5 years or more (Santen et al. 1990; Jordan and Murphy 1990). A former concern with the prolonged use of tamoxifen was a further acceleration of age-related bone loss. However, studies of postmenopausal patients treated with tamoxifen for breast cancer have not demonstrated increased bone resorption and most studies suggest that tamoxifen has a protective effect on the skeleton (Fentiman et al. 1989; Gotfredson et al. 1984; Kristensen et al. 1989; Love et al. 1988). Furthermore, tamoxifen was found to increase high density lipoprotein cholesterol, sex hormone-binding globulin, and T_4-binding globulin while simultaneously decreasing low density lipoprotein cholesterol (Kristensen et al. 1989; Love et al. 1988), suggesting that it has an estrogen-like action on the liver.

The effects of tamoxifen on bone and mineral metabolism have not been investigated in premenopausal women. It should be emphasized that the beneficial effects of tamoxifen observed in postmenopausal women with low endogenous levels of estrogen may not occur in premenopausal women with high circulating levels of the hormone. Kalu et al. (1991) demonstrated that tamoxifen partially antagonizes the skeletal effects of pharmacological doses of estrogen to ovariectomized rats, a result that was confirmed by Moon et al. (1991).

4.3.2 Mechanisms of Action of Tamoxifen

The mechanisms of the tissue-selective actions of tamoxifen and other estrogen analogs are poorly understood and likely complex. Tamoxifen is reported to bind to the estrogen receptor with lower affinity than estrogen and possibly at regions distinct from the estradiol binding site (Tate et al. 1984a, b; Martin et al. 1988). There is evidence that anti-

---

Fig. 3. *(cont.).* Nevertheless, the results suggest that tamoxifen acts primarily as an estrogen agonist to prevent osteopenia in the ovariectomized rat. Though tamoxifen antagonizes pharmacological levels of administered estrogen (Moon et al. 1991; Kalu et al. 1991), it appears to only weakly antagonize endogenous levels of the hormone

estrogen–estrogen receptor complexes bind to DNA but are unable to stimulate transcription (Webster et al. 1988). Unfortunately, there have been few studies to determine whether the tissue-selective actions of estrogen analogs are related to tissue-specific differences in the binding of the ligand–receptor complex to estrogen-responsive elements. Initial studies suggest, however, that estrogen analogs interact differently than estrogen with gene-activating domains on the estrogen receptor (Tzukerman et al. 1994).

Indirect actions of estrogen analogs could potentially influence individual tissue responses to analogs. Tamoxifen, for example, induces an increase in the circulating level of estrogen (Sherman et al. 1979; Rose and Davis 1980; Manni and Pearson 1980; Tajima and Fukushima 1983; Dnistrian et al. 1985) in premenopausal women and may reduce the level of free estradiol in postmenopausal women (Sakai et al. 1978; Szamel et al. 1986; Jordan et al. 1987a). It also reduces the circulating levels of insulin-like growth factor-1 which in of itself might mediate selected tissue-selective responses (Colletti et al. 1989). Finally, metabolites of tamoxifen which are biologically active may be generated(Jordan 1982) and may accumulate in specific tissues (Robinson et al. 1991).

4.3.3 Bone Cell-Selective Actions of Estrogen Analogs

As mentioned, bone formation and bone resorption are coupled during bone remodeling. This coupling is imperfect in estrogen-depleted individuals such that there is net bone loss. To restore bone to an osteoporotic skeleton, it is insufficient to reduce bone turnover, the principal effect of estrogen replacement therapy. To restore bone it is necessary to improve bone balance by establishing a *net* increase in bone formation. Theoretically, estrogen analogs may prove more effective in accomplishing this goal than natural estrogens. It may prove possible to identify estrogen analogs with selective actions on bone forming and bone resorbing cells. Estrogen receptors are located on both osteoclasts and osteoblasts (Oursler et al. 1991). It may prove possible to take advantage of the presence of estrogen receptors and presumed unique actions of the hormone on the two cell lineages to partially dissociate the actions of estrogen on bone formation and bone resorption. While it is not yet possible to restore lost bone in animal models for estrogen deficiency with estrogen analogs, there is evidence that some analogs have actions on the skeleton which differ from estrogen. For example, raloxifene preserved cancellous bone in ovariectomized rats and decreased indices of bone resorption to a greater extent than bone formation (Evans et al. 1994). More recently, clomiphene was shown in the author's laboratory to not only preserve cancellous bone in ovariectomized rats, but to even increase mean trabecular thickness. This finding contrasts with estradiol-treated rats in which no change in trabecular thickness was observed. Although not definitive, these observations suggest that tissue and bone cell-selective estrogen analogs have more potential to improve the efficacy of hormone replacement therapy than natural estrogens.

4.4 Summary

Estrogen analogs which are partial estrogen agonists/antagonists can have organ-selective activities. It has proven possible to identify analogs which are primarily estrogen agonists on bone and liver and antagonists on reproductive tissues. These organ-selective estrogen analogs may be superior to natural estrogens for hormone replacement therapy by reducing side effects associated with estrogen action on reproductive tissues. If new analogs are identified which can distinguish between bone forming and bone resorbing cells it may prove possible to improve the efficacy of estrogen replacement therapy.

References

Aitken JM, Hart DM, Lindsay R (1973) Estrogen replacement therapy for prevention of osteoporosis after oophorectomy. Br Med J 3:515–518

Arnold JS (1973) Amount and quality of trabecular bone in osteoporotic and vertebral fractures. Clin Endocrinol Metab 2:221–238

Beall PT, Misra LK, Young RL, Spjut HJ, Evans HJ, LeBlanc A (1984) Clomiphene protects against osteoporosis in the mature ovariectomized rat. Calcif Tissue Int 36:123–125

Bellido T, Girasole G, Passeri G, Yu X-P, Mocharla H, Jilka RL, Notides A, Manolagas SC (1993) Demonstration of estrogen and vitamin D receptors in bone marrow-derived stromal cells: up-regulation of the estrogen receptor by 1,25-dihydroxyvitamin-D_3. Endocrinology 133:553–562

Caputo CB, Meadows D, Raisz LG (1976) Failure of estrogens and androgens to inhibit bone resorption in tissue culture. Endocrinology 98:1065–1068

Chen TL, Feldman D (1978) Distinction between alpha-fetoprotein and intracellular estrogen receptors: evidence against the presence of estradiol receptors in rat bone. Endocrinology 102:236–244

Chow J, Tobias JH, Colston KW, Chambers TJ (1992) Estrogen maintains trabecular bone volume in rats not only by suppression of bone resorption but also by stimulation of bone formation. J Clin Invest 89:74–78

Colletti RB, Roberts JD, Devlin JT, Copeland KC (1989) Effect of tamoxifen on plasma insulin-like growth factor I in patients with breast cancer. Cancer Res 49:1882–1884

Cummings SR, Kelsey JL, Nevitt MC, O'Dowd KJ (1985) Epidemiology of osteoporosis and osteoporotic fractures. Epidemiol Rev 7:178–208

Dnistrian AM, Greenberg EJ, Dillon HJ, Hakes TB, Fracchia AA, Schwartz MK (1985) Chemohormonal therapy and endocrine function in breast cancer patients. Cancer 56:63–70

Dukes M, Chester R, Yarwood L, Wakeling AE (1994) Effects of non-steroidal pure antiestrogen, ZM 189,154, on oestrogen target organs of the rat including bones. J Endocrinol 141:335–341

Eastell R, Delmas PD, Hodgson SF, Eriksen EF, Mann KG, Riggs BL (1988) Bone formation rate in older normal women: concurrent assessment with bone histomorphometry, calcium kinetics, and biochemical markers. J Clin Endocrinol Metab 67:741–748

Eriksen EF, Colvard DS, Berg NJ, Graham ML, Mann KG, Spelsberg TC, Riggs BL (1988) Evidence of estrogen receptors in normal human osteoblast-like cells. Science 241:84–86

Evans GL, Bryant HU, Magee D, Sato M, Turner RT (1994) Raloxifene is a tissue specific estrogen agonist which prevents osteopenia in ovariectomized rats. Endocrinology 134:2283–2288

Feldman S, Minne HW, Parviz S, Pfeifer M, Lempert VG, Bauss F, Zeigler R (1989) Antiestrogen and antiandrogen administration reduce bone mass in the rat. J Bone Miner Res 7:245–254

Fentiman CM, Rodin A, Murby B, Fogelman I (1989) Bone mineral content of women receiving tamoxifen for mastalgia. Br J Cancer 60:262–264

Frenay M, Milano G, Formento JL, Francoual M, Moll JL, Namer M (1991) Oestrogen and progesterone receptor status in bone biopsy specimens from patients with breast cancer. Eur J Cancer 27:115–118

Frost HM (1963) Dynamics of bone remodeling. In: Frost JM (ed) Bone biodynamics. Little Brown, Boston, pp 1–168

Gallagher A, Chambers TJ, Tobias JH (1993) The estrogen antagonist ICI 182,780 reduces cancellous bone volume in female rats. Endocrinology 133:2787–2791

Garn SM, Rohmann CG, Wagner B (1967) Bone loss as a general phenomenon in man. Fed Proc 26:1729–1736

Genant HK, Baylink DJ, Gallagher JC (1989) Estrogens in the prevention of osteoporosis in postmenopausal women. Am J Obstet Gynecol 161:1842–1846

Gotfredson A, Christiansen C, Palshof T (1984) The effect of tamoxifen on bone mineral content in premenopausal women with breast cancer. Cancer 53:853–857

Hazzard WR (1989) Estrogen replacement and cardiovascular disease: serum lipids and blood pressure effects. Am J Obstet Gynecol 161:1847–1853

Henderson BE (1989) The cancer question: an overview of recent epidemiologic and retrospective data. Am J Obstet Gynecol 161:1859–1864

Jordan VC (1982) Metabolites of tamoxifen in animals and man: identification, pharmacology and significance. Breast Cancer Res Treat 2:123–138

Jordan VC, Murphy CS (1990) Endocrine pharmacology of antiestrogens as antitumor agents. Endocr Rev 11:578–610

Jordan VC, Fritz NF, Tormey DC (1987a) Long-term adjuvant therapy with tamoxifen: effect on sex hormone binding globulin and antithrombin III. Cancer Res 47:4517–4519

Jordan VC, Phelps E, Lindgren JV (1987b) Effects of antiestrogens on bone in castrated and intact female rats. Breast Cancer Res Treat 10:31–35

Kalu DN, Liu CC, Hardin RR, Hollis BW (1989) The aged rat model of ovarian hormone deficiency bone loss. Endocrinology 124:7–16

Kalu DN, Salerno E, Liu CC, Echon R, Ray M, Garza-Zapata M (1991) A comparative study of the actions of tamoxifen, estrogen, and progesterone in the ovariectomized rat. Bone Miner 15:109–123

Komm BS, Terpening CM, Benz DJ, Graeme KA, Gallegos A, Kore M, Greene GL, O'Malley BW, Haussler MR (1988) Estrogen binding, receptor mRNA and biologic response in osteoblast-like osteosarcoma cells. Science 241:81–84

Kristensen B, Moridsen HT, Holmegaard SN, Transbol I (1989) Amelioration of postmenopausal primary hyperparathyroidism during adjuvant tamoxifen for breast cancer. Cancer 64:1965–1967

Love RR, Mazeus RB, Tormay DC, Barden HS, Newcomb PA, Hordan VC (1988) Bone mineral density in women with breast cancer treated with adjuvant tamoxifen for two years. Breast Cancer Res Treatment 12:297–301

Mack TM, Ross RK (1989) Risks and benefits of long-term treatment with estrogens. Schweiz Med Wochenschr 119:1811–1820

Manni A, Pearson OH (1980) Antiestrogen-induced remission in premenopausal women with stage IV breast cancer: effects on ovarian function. Cancer Treat Rep 64:779–785

Martin PM, Berthois Y, Jensen EV (1988) Binding of antiestrogens exposes an occult antigenic determinant in the human estrogen receptor. Proc Natl Acad Sci USA 85:2533–2537

Miller SC, Wronski TJ (1993) Long-term osteopenic changes in cancellous bone structure in ovariectomized rats. Anat Rec 236:433–441

Moon L, Wakley GK, Turner RT (1991) Dose and ovary dependent effects of tamoxifen on bone balance in tibia of maturing rats. Endocrinology 129:1568–1574

Oursler M, Osdoby P, Pyfferoen J, Riggs BL, Spelsberg TC (1991) Avian osteoclasts as estrogen target cells. Proc Natl Acad Sci USA 88:6613–6617

Riggs BL, Wahner HW, Dunn WL, Mazess RB, Offord KP, Melton LJ (1981) Differential changes in bone mineral density of the appendicular and axial skeleton with aging. J Clin Invest 67:328–335

Riggs BL, Wahner HW, Melton LJ III, Richelson LS, Judd HL, Offord KP (1986) Rates of bone loss in the appendicular and axial skeletons of women. J Clin Invest 77:1487–1491

Robinson SP, Langan-Fahey SM, Johnson DA, Jordan VC (1991) Metabolites, pharmacodynamics and pharmacokinetics of tamoxifen in rats and mice compared to the breast cancer patient. Drug Metab Dispos 19:36–43

Rose DP, Davis TE (1980) Effects of adjuvant chemohormonal therapy on the ovarian and adrenal function of breast cancer patients. Cancer Res 40:4043–4047

Sakai F, Cheix F, Clavel M, Colon J, Mayer M, Pommatau E, Saez S (1978) Increases in steroid binding globulins induced by tamoxifen in patients with carcinoma of the breast. J Endocrinol 76:219–226

Santen RJ, Manni A, Harvey H, Redmond C (1990) Endocrine treatment of breast cancer in women. Endocr Rev 11:221–265

Sherman BM, Chapler FK, Crickard K, Wycoff D (1979) Endocrine consequences of continuous antiestrogen therapy with tamoxifen in premenopausal women. J Clin Invest 64:398–404

Somjen D, Weisman Y, Harrel A, Berger E, Kaye AM (1989) Direct and sex-specific stimulation by sex steroids of creatine kinase activity and DNA synthesis in rat bone. Proc Natl Acad Sci USA 86:3361–3365

Stewart PJ, Stern PH (1986) Effect of the antiestrogens tamoxifen and clomiphene on bone resorption in vitro. Endocrinology 118:125–131

Szamel I, Vincze B, Hindy I, Herman I, Borvendeg J, Eckhardt S (1986) Hormonal changes during a prolonged tamoxifen treatment in patients with advanced breast cancer. Oncology 43:7–11

Tajima C, Fukushima T (1983) Endocrine profiles in tamoxifen-induced ovulatory cycles. Fertil Steril 40:23–30

Tate AC, Greene GL, DeSombre ER, Jensen EV, Jordan VC (1984a) Differences between estrogen and antiestrogen-estrogen receptor complexes from human breast tumors identified with an antibody raised against the estrogen receptor. Cancer Res 44:1012–1018

Tate AC, Lieberman ME, Jordan VC (1984b) The inhibition of prolactin synthesis in GH3 rat pituitary tumor cells by monohydroxytamoxifen is associated with changes in the properties of the estrogen receptor. J Steroid Biochem 20:391–395

Turner RT, Vandersteenhoven JJ, Bell NH (1987a) The effects of ovariectomy and 17 beta estradiol on cortical bone histomorphometry in growing rats. J Bone Miner Res 2:115–122

Turner RT, Wakley GK, Hannon KS, Bell NH (1987b) Tamoxifen prevents the altered bone turnover resulting from ovarian hormone deficiency. J Bone Miner Res 2:449–456

Turner RT, Wakley GK, Hannon KS, Bell NH (1988) Tamoxifen inhibits osteoclast mediated resorption of trabecular bone in ovarian hormone deficient rats. Endocrinology 122:1146–1150

Turner RT, Hannon KS, Demers L, Buchanan J, Bell NH (1989) Differential effects of gonadal function on bone histomorphometry in male and female rats. J Bone Miner Res 4:557–563

Turner RT, Colvard DS, Spelsberg TC (1990) Estrogen inhibition of periosteal bone formation in rat long bones: down regulation of gene expression for bone matrix proteins. Endocrinology 127:1346–1351

Turner RT, Bell NH, Gay CV (1993a) Estrogen receptors and bone: evidence that estrogen-binding sites are present in bone cells and mediate medullary bone formation in Japanese quail. Poultry Sci 72:728–740

Turner RT, Evans GL, Wakley GK (1993b) The mechanism of action of estrogen on cancellous bone balance in tibial of ovariectomized growing rats: inhibition of indices of formation and resorption. J Bone Miner Res 8:359–366

Turner RT, Evans GL, Wakley GK (1994a) Reduced chondroclast differentiation results in increased cancellous bone volume in estrogen-treated growing rats. Endocrinology 134:461–466

Turner RT, Riggs BL, Spelsberg TC (1994b) Skeletal effects of estrogen. Endocr Rev 15:275–300

Tzukerman MT, Esty A, Santiso-Mere D, Danielian P, Parker MG, Stein RB, Pike JW, McDonnell DP (1994) Human estrogen receptor transactivational capacity is determined by both cellular and promotor context and mediated by two functionally distinct intramolecular regions. Mol Endocrinol 8:21–30

Wakley GK, Baum RL, Hannon KS, Turner RT (1988) Tamoxifen treatment reduces osteopenia induced by immobilization in the rat. Calcif Tissue Int 43:383–388

Wakley GK, Shutte DE, Hannon KS, Turner RT (1991) The effects of castration and androgen replacement therapy on bone: a histomorphometric study in the rat. J Bone Miner Res 6:325–330

Waters DJ, Caywood DD, Turner RT (1991) Effect of tamoxifen citrate on canine immobilization (disuse) osteoporosis. Vet Surg 20:392–396

Webster NJCT, Green S, Jin SR, Chambon P (1988) The hormone-binding domains of the estrogen and glucocorticoid receptors contain an inducible transcription activation function. Cell 54:199–207

Westerlind KC, Evans GL, Wronski TJ, Bell NH, Turner RT (1995) Estrogen regulates the rate of bone turnover but bone balance is determined by mechanical strain. Presented at the XIIth international conference on calcium regulating hormones, Melbourne, Australia, 14–19 February

Westerlind KC, Wakley GK, Evans GL, Turner RT (1993) Estrogen does not increase bone formation in growing rats. Endocrinology 133:2924–2934

Williams DC, Paul DC, Black LJ (1991) Effects of estrogen and tamoxifen on serum osteocalcin levels in ovariectomized rats. Bone Miner 14:205–220

Wronski TJ, Cintron M, Doherty SL, Dann LM (1988) Estrogen treatment prevents osteopenia and depresses bone turnover in ovariectomized rats. Endocrinology 123:681–686

Wronski TJ, Lowry PL, Walsh CC, Ignaszewski LA (1985) Skeletal alterations in ovariectomized rats. Calcif Tissue Int 37:324–328

Wronski TJ, Walsh CC, Ignaszewski LA (1986) Histological evidence for osteopenia and increased bone turnover in ovariectomized rats. Bone 7:119–123

Yamamoto TT, Rodan GA (1990) Direct effects of 17β-estradiol on trabecular bone in ovariectomized rats. Proc Natl Acad Sci USA 87:2172–2176

Yoshioka T, Sato B, Matsumoto K, Ono K (1980) Steroid receptors in osteoblasts. Clin Orthop 148:297–303

5 Progestin Regulation
of Cell Proliferation in the Breast
and Endometrium

R. L. Sutherland, C. S. L. Lee, A. L. Cornish,
and E. A. Musgrove

5.1	Introduction	85
5.2	Effects of Progestins on Cell Proliferation In Vivo	86
5.2.1	Effects in the Uterus	86
5.2.2	Effects in the Mammary Gland	89
5.2.3	Differential Effects in the Breast and Endometrium	92
5.3	Breast Cancer Cells as an In Vitro Model for Progestin Action	93
5.4	Mechanisms of Cell Cycle Control	96
5.5	Progestin Effects on CDK Function	98
5.6	Conclusions	101
References		103

5.1 Introduction

The regulation of growth and development of female sex organs involves a balance between the actions of the two major female sex steroid
hormones, oestradiol and progesterone. While oestrogen, acting in concert with other hormones and growth factors, appears to be the main
drive to proliferation in these tissues, progesterone has two principal
functions in normal mammalian physiology. First, progesterone is involved in preparing the uterus for implantation of the fertilized ovum
and making nutrients available for its subsequent development. Second,

progesterone causes the glandular elements of the mammary gland to grow and develop into secretory epithelium with the ultimate effect of acting in concert with other hormones, particularly prolactin, to facilitate milk production. In simplistic terms, progesterone might be seen as the "differentiating" female sex steroid which inhibits the "proliferative" actions of oestrogen and directs the tissue towards its normal differentiated function. However, progesterone is not always "antiproliferative" and in some tissues induces proliferative responses of its own. In the case of the induction of stromal proliferation in the uterus this represents a corollary of its primary function in facilitating implantation; in the case of its stimulation of lobuloalveolar development in the mammary gland, such an action is a requirement for subsequent lactation, the ultimate differentiated function of this organ.

The data reported here focus on the effects of progesterone and synthetic progestins on cell proliferation, an area of cell biology that has not been widely studied from a mechanistic viewpoint (reviewed in Clarke and Sutherland 1990, 1993). Emphasis is placed on progestin effects in the endometrium and the mammary gland, since the increasing pharmacological use of progestins in oral contraceptives and hormone replacement therapies raises important questions of potential adverse side effects, including the possibility that progestins may contribute to breast cancer risk.

5.2 Effects of Progestins on Cell Proliferation In Vivo

5.2.1 Effects in the Uterus

The mouse and rat uterus are the most widely studied systems for investigating the effects of oestrogen and progesterone on cell proliferation in vivo (reviewed in Clarke and Sutherland 1990). In these models oestradiol alone causes a major mitogenic response in epithelial cells but not in the connective tissue stroma, while progesterone significantly alters the proliferative response to oestradiol (Martin and Finn 1971). Pretreatment with progesterone completely inhibits oestrogen-induced epithelial cell proliferation while sensitizing the stromal cells to respond to oestradiol with increased mitoses. The progestin-induced switch in proliferation from epithelium to stroma is an essential prerequisite for implantation and decidualization in the mouse and rat, emphasizing the differentiation-inducing role of progesterone. It is also of major interest that in the mouse model, while progesterone inhibits oestrogen-induced mitosis in the epithelium, it acts synergistically with oestrogen to stimulate stromal proliferation. These data highlight the cell specificity of progesterone action and its ability to both stimulate and inhibit cell proliferation in different cell types.

These early studies also provided insight into the cell kinetic basis of progestin antagonism of oestrogen-induced epithelial cell proliferation. While prior administration of progesterone or simultaneous administration of oestrogen and progesterone completely blocked the oestrogenic response, no inhibition was seen when progesterone was given as shortly as 2.5 h after oestradiol (Das and Martin 1973; Martin et al. 1973). Since oestrogen induces synchronous progression of resting G_0 phase uterine epithelial cells through the cell cycle, progesterone was proposed to inhibit cell proliferation by an action early in G_1 phase and was without effect on progression through late G_1, S and G_2 phases, a conclusion subsequently confirmed by more mechanistic studies with breast cancer cells in culture (see below). The inhibition of epithelial cell proliferation by progesterone is associated with morphological changes characteristic of epithelial cell differentiation.

In the progesterone-treated stroma, a single injection of oestradiol resulted in synchronous entry of 30%–40% of stromal cells into S phase whilst a second injection of oestradiol produced no further effect. It thus appears that progesterone stimulates resting stromal cells to enter the cell cycle where oestrogen accelerates their passage through a single round of replication by shortening G_1 phase. This replication is thought to be a prerequisite for the differentiation of stromal cells into decidual cells and their withdrawal from the cell cycle.

These observations in the mouse uterus identified critical issues for consideration in defining the molecular basis of progestin effects on cell proliferation. In particular, the complexity of the response was illustrated by observations on cell type specificity of progestin responsiveness, i.e. between epithelium and stroma, and the dependence of the proliferative response on the temporal relationship between administration of oestradiol and progesterone. Perhaps more importantly from a mechanistic viewpoint, the data on cell cycle kinetics illustrated the inhibition of cell cycle progression in early G_1 phase in epithelial cells

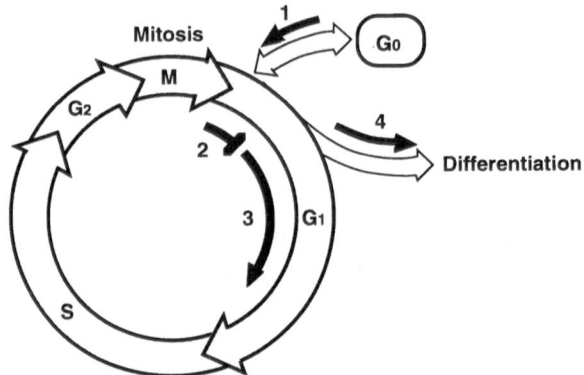

Fig. 1. Cell cycle phase specific effects of progestins. The four phases of the cell cycle: G_1, DNA synthesis or S phase, G_2 and mitosis (M) are illustrated. Cells can leave the cell cycle and enter a resting or G_0 phase which allows re-entry to the cell cycle. Alternatively, cells can leave the cell cycle to enter an irreversible program of cell differentiation. Examples of the effects of progestins are shown, i.e. *1*, progesterone induction of resting G_0 cells into the cell cycle in mouse uterine stroma; *2*, inhibition of cell cycle progression in early G_1 phase in uterine and mammary epithelial cells; *3*, accelerated progression of cells through G_1 phase in mammary carcinoma cells and *4*, terminal differentiation in the uterus and mammary gland

and activation of G_0 stromal cells (Fig. 1) and identified that progestins have both stimulatory and inhibitory effects on target cell proliferation (Martin and Finn 1971).

Several of the principles of progesterone action identified in experimental animal models were subsequently confirmed in the human endometrium (Creasy et al. 1992). The cyclical histological changes in the endometrium during the human menstrual cycle are well correlated with changes in circulating concentrations of oestrogen and progesterone (Fig. 2). Cyclical changes in proliferation occur, with DNA synthesis being maximal around the time of ovulation then decreasing to low levels until the end of the cycle. Cell proliferation in both the glandular epithelium and stromal elements increases as the serum oestrogen concentration increases, while the rise in serum progesterone in the postovulatory phase results in the disappearance of mitoses in both epithelial and stromal cells, indicating that progesterone is able to in-

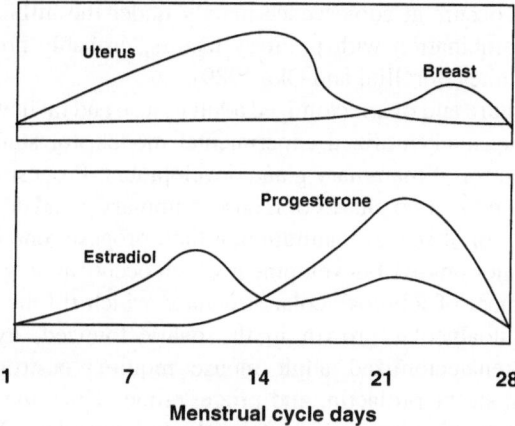

Fig. 2. Changes in uterine and breast epithelial mitoses (*top*) and serum estradiol and progesterone levels (*bottom*) during the menstrual cycle. Redrawn from Clarke and Sutherland (1990)

hibit oestrogen-induced proliferation in these cells. The hypothesis that progesterone inhibited oestrogen-mediated DNA synthesis was confirmed by progestin inhibition in vivo of oestrogenized postmenopausal endometrial proliferation (Whitehead et al. 1981). The inhibition of mitosis was accompanied by induction of glandular secretory activity and demonstrated the ability of progestins to promote differentiated function in postmenopausal women on oestrogen therapy. Progestins also inhibit the growth of endometrial carcinoma tissue, confirming their predominant inhibitory effect on epithelial cell proliferation in the mammalian uterus (Creasy et al. 1992).

5.2.2 Effects in the Mammary Gland

Although mammary epithelial development involves complex interactions between a number of different hormones which vary between species and at different stages of development, the principal role of progesterone is in promoting lobuloalveolar development in the adult gland. Progesterone does not appear to be necessary for ductal develop-

ment, which occurs at adolescence mainly under the influence of oestrogen in combination with pituitary factors, probably prolactin and
growth hormone (Borellini and Oka 1989).

The immature and ovariectomized adult mouse and rat have provided
the most extensively utilized experimental models for studies on the
hormonal control of mammary gland development. Progesterone is not
involved in the first two phases of mouse mammary gland development,
i.e. prenatally or at sexual maturation; rather, progesterone is involved
in the last major phase of development which occurs at pregnancy with
the development of lobuloalveolar structures which fill the interductal
spaces. Lobuloalveolar growth in the ovariectomized, hypophysec-
tomized, adrenalectomized adult mouse requires oestrogen, either
growth hormone or prolactin, and progesterone. Thus, the major difference between stimulation of ductal and lobuloalveolar cell proliferation during development is the additional requirement for progesterone
in the latter cell type (Clarke and Sutherland 1990).

In the ductal epithelium of the fully mature gland, progesterone and
oestradiol alone both stimulate proliferation, though progesterone is
considerably more effective. Administration of the two steroids together
results in a marked synergistic, although transient, effect which was
attributed in part to the ability of oestrogen to increase progesterone
receptor levels and progestin responsiveness. These studies were interpreted as evidence that in the mature mouse progesterone, rather than
oestrogen, has a major role in promoting epithelial cell proliferation
(Haslam 1988). This adds another degree of complexity to progestin
action by providing evidence for developmental differences in the hormonal responsiveness of mammary epithelial cells, i.e. progestins are
not required for ductal epithelial cell proliferation during development
but have a major influence on these cells in the mature gland. Further
complexity is evident from studies in rats, demonstrating differential
responsiveness to progestins within the terminal ductal structures of the
mammary gland. Although ovariectomy reduced DNA synthesis in
terminal end buds, alveolar buds and terminal ducts, indicating a dependence of cell proliferation on ovarian hormones, progestins inhibited
DNA synthesis in the latter two structures only; in the terminal end buds
progestins had no effect or stimulated proliferation (Russo and Russo
1991).

The human breast responds to the fluctuations in serum hormone levels during the menstrual cycle with cyclical changes in breast volume and cellular morphology (see Laidlaw et al. 1995 for references). Contributing to these effects are cyclical changes in the mitotic activity of the epithelium; the highest proliferative activity was noted in the intralobular terminal ducts. Studies in which mitoses were evaluated histologically showed that both mitosis and cell loss through apoptosis varied in a cyclical manner during the menstrual cycle, with mitoses being maximal on days 23–26, (Fig. 2), although there was variation in the ability to detect mitoses in the early follicular phase. The stimulus for this wave of epithelial cell proliferation in the late secretory phase of the cycle is presently unknown. The effect coincides with, or immediately follows, a rise in the serum concentrations of both oestrogen and progesterone (Fig. 2). The inference can thus be drawn that either or both of these hormones, directly or indirectly, may be responsible, perhaps in an analogous manner to the synergism seen in the mature mouse mammary gland. However, extrapolation from animal models to humans must be interpreted with caution, given that at least four distinct types of increasingly more differentiated lobular structures have been identified in the female breast and their ratios vary significantly with parity and age (Russo and Russo 1993). Thus, in humans the effects of exogenous progestins on breast epithelial cell proliferation are likely to differ with age and functional development of the gland.

The roles of oestrogen and/or progesterone in mediating breast epithelial DNA synthesis in humans remains controversial but the issue has been examined using in vitro techniques such as organ culture, transplantation of normal human breast tissue into nude mice and primary cell culture. The consensus to emerge from these studies is that oestrogen is capable of stimulating breast epithelial growth both in vitro and in vivo. Responses to progesterone, however, were variable and no clear consensus on a stimulatory role for progesterone or an ability to inhibit the oestrogen-mediated effect has yet emerged, although progestin treatment inhibits the proliferation of steroid-responsive breast epithelial cells in short-term culture (Gompel et al. 1986). A more recent study has addressed the issue of which ovarian steroids stimulate normal human mammary epithelial cell proliferation by implantation of normal breast tissue into nude mice followed by treatment with oestradiol and progesterone. Oestradiol stimulated the thymidine labelling index while progesterone had no effect either when administered alone or in combination with oestradiol, despite the presence of oestrogen-inducible progesterone receptors in this tissue (Laidlaw et al. 1995). Breast tissue is composed of ductal, lobular and alveolar epithelial elements, encased by myoepithelial cells, and surrounded by stroma, and it is likely that these different cellular elements will respond differently to progestins and other growth regulators. Further experimentation in this area is critical to understanding the role of progesterone in the control of cell proliferation in the human breast. However, the fact that progesterone failed to stimulate proliferation in the normal breast and that synthetic progestins are effective agents in the treatment of metastatic breast and endometrial carcinoma argues against a major proliferative role for progestins in breast epithelial cells. Clinical data relating objective responses following progestin treatment to the presence of PR support a direct receptor-mediated effect on breast carcinoma cells (Sedlacek and Horwitz 1984), a conclusion supported by studies in vitro.

5.2.3 Differential Effects in the Breast and Endometrium

The demonstrated antiproliferative effects of progestins in the endometrium of a range of animal models plus the clinical efficacy of progestins in both opposing the carcinogenic effects of oestrogens in the uterus and the treatment of established endometrial carcinoma (Creasy et al. 1992; Hulka et al. 1994) support a primarily inhibitory role for progestins in the control of cell proliferation in the uterus. These data from the endometrium were initially extrapolated to the breast in the absence of direct experimental data on this tissue. However, more recent studies relating breast epithelial cell mitosis to later stages of the menstrual cycle than endometrial mitosis has radically changed views on the role of progestins in the control of breast cell proliferation. Several authors have proposed that progestins have a primary mitogenic role in the human breast although direct evidence for this is lacking, as evidenced by studies with normal breast epithelium both in vivo (Laidlaw et al. 1995) and in vitro (Gompel et al. 1986). Nonetheless, the association of increased breast epithelial mitosis with increased oestradiol and progesterone levels in the luteal phase of the menstrual cycle has prompted the hypothesis that this increased rate of cell proliferation could contribute to breast cancer development by increasing the possibility of oncogenic mutations (Henderson et al. 1991). Thus, in the absence of direct evidence, it is hypothesized that there are apparently opposing effects of progesterone in the endometrium and breast, i.e. in the former progesterone is growth inhibitory and protective of cancer development whereas in the latter it is stimulatory and carcinogenic.

There are, however, alternative explanations for the data accumulated to date. For example, increased proliferation as measured by the thymidine labelling index identifies cells synthesizing DNA but does not provide evidence for a sustained proliferative effect on a stem cell population which would be necessary to increase the probability of mutation and carcinogenesis. An alternative explanation is that the cells which undergo mitosis do so on their path to terminal differentiation and exit from the cell cycle. If this were the end result of the progestin effect as observed in several of the experimental models discussed above, progestin administration would be expected to have little effect on breast cancer risk.

These and other considerations prompted us to study in more detail the effects of progestins on target cell proliferation in a well defined experimental system in vitro.

5.3 Breast Cancer Cells as an In Vitro Model for Progestin Action

Steroid-responsive breast cancer cells are a widely-used model for studying steroid hormone action and in this role have provided a number of insights into progestin action on cell proliferation. Early studies, including those from this laboratory, showed predominantly inhibitory effects in cells stimulated to proliferate with fetal calf serum and oestrogen (Vignon et al. 1983; Horwitz and Freidenberg 1985; Sutherland et al. 1988) although stimulatory responses were sporadically reported. However, a more detailed analysis of the effects of progestins in T-47D cells growing at suboptimal growth rates under defined culture conditions (Musgrove et al. 1991) revealed both stimulation and inhibition, providing evidence for two distinct effects of progestins on cell cycle progression within the one cell type. Together these effects resulted in a biphasic change in the rate of cell cycle progression, consisting of an

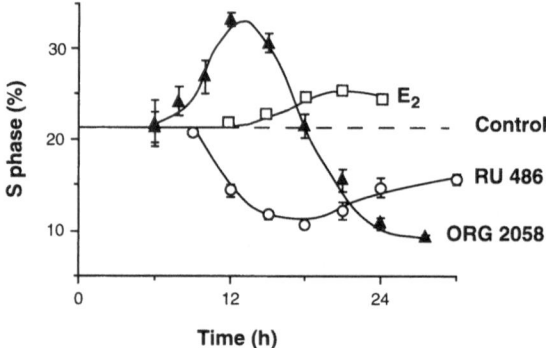

Fig. 3. Comparison of temporal changes in %S phase after progestin, oe-strogen or antiprogestin treatment. T-47D human breast cancer cells prolif-erating exponentially in insulin-supplemented serum free medium were treated with either synthetic progestin (ORG 2058), oestradiol (E_2) or antiprogestin (RU 486) and the proportion of cells in S phase determined at intervals. Data have been redrawn from Musgrove et al. (1991) and Musgrove and Sutherland (1993)

initial transient acceleration through G_1 and subsequent increase in the %S phase, followed by cell cycle arrest and growth inhibition accompa-nied by a decrease in the %S phase (Figs. 1,3). The decreased %S phase is maintained so that the predominant effect is long-term growth inhibi-tion (Fig. 3).

The two effects could not be separated on the basis of progestin concentration-dependence, or concentration- or time-dependence of RU 486 antagonism: in each case any reduction in magnitude of one compo-nent of the response was accompanied by a decrease in the magnitude of the other (Musgrove et al. 1991). Furthermore, a range of structurally distinct synthetic progestins each exhibited a clearly biphasic response (Musgrove et al. 1991). Since many synthetic progestins in therapeutic use have appreciable affinity for receptors other than the progesterone receptor, for example the glucocorticoid, androgen and oestrogen recep-tors, the effects of ligands with specificity for particular receptors were compared with the biphasic response of progestins. Androgens and the synthetic glucocorticoid dexamethasone have minimal effects in this experimental model (Sutherland et al. 1988; Musgrove et al. 1991), which may be due in part to the low level of expression of the cognate receptors relative to the high concentration of progesterone receptors in T-47D cells (Hall et al. 1990, 1992). The effects of natural oestrogens and progestin antagonists are more pronounced but are distinct from either the stimulatory or inhibitory effects of progestins. Oestrogens are mitogenic, leading to an increase in the %S phase which is of smaller magnitude than the increase following progestin treatment and first apparent 4–6 h after progestin stimulation (Fig. 3). Progestin antagonists, like progestins, inhibit breast cancer cell proliferation, but the resulting decrease in %S phase becomes apparent 3–6 h before progestin-induced decreases in %S phase (Fig. 3). Indeed, progestin stimulation and progestin antagonist inhibition become apparent over a similar time course, suggesting actions at a similar portion of G_1 phase. An attractive hypothesis is that these effects are mediated by opposing actions on the same targets, rather than that the growth inhibition by progestin antagonists indicates progestin agonist activity. In summary, the biphasic effects of progestins in T-47D cells are mediated largely, if not entirely, through the progesterone receptor, thus providing a model system ideal for studying these effects in isolation from those mediated via other members of the steroid hormone receptor family.

The proportion of the total cell population which is accelerated into S phase following progestin treatment depends on the growth rate of the cells prior to treatment (Musgrove et al. 1991). This suggests that rather than stimulating growth-arrested cells to re-enter the cell cycle as progestins do in uterine stromal cells, progestins accelerated breast cancer cells which were already "in cycle". In turn, this implies that a direct or indirect target of progestin action is a rate-limiting step in cell cycle progression, which appears to be in mid-G_1 phase (Fig. 4). Since cell numbers approximately double before the reduced proliferation rate becomes apparent, most of the cell population is capable of completing a round of replication before ultimately arresting in G_1 phase, indicating that the inhibitory effect occurs in early G_1 phase (Sutherland et al. 1988; Musgrove et al. 1991) (Fig. 4). Thus, the two effects of progestins target temporally distinct processes within G_1. One possibility is that these effects are mediated by two independent mechanisms. However, since in a number of tissues progestins can be viewed as inducers of differentiation (Clarke and Sutherland 1990), growth arrest might be a consequence of the initiation of a differentiation programme. The transient increase in cell cycle progression might then arise from a necessity

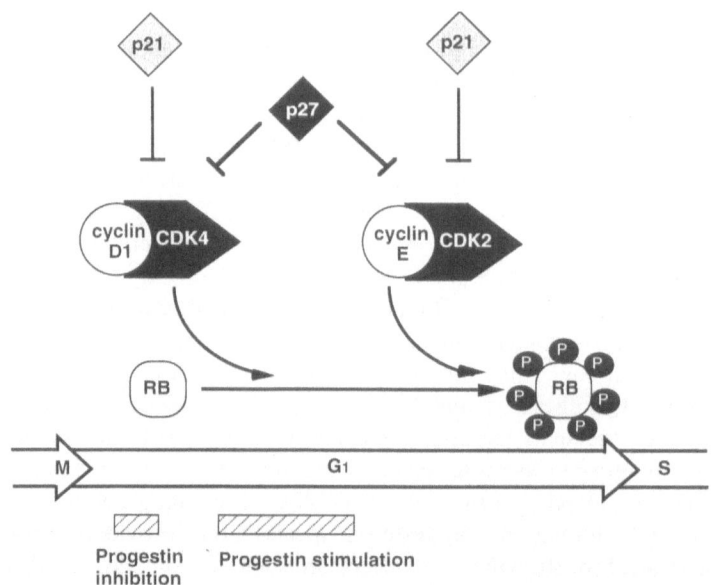

Fig. 4. Control of G₁ phase progression in breast cancer cells. Following mitosis (*M*) sequential activation of cyclin D1/Cdk4 and cyclin E/Cdk2 leads to phosphorylation (*P*) of the tumour suppressor gene product *RB* followed by entry into S phase. These cyclin/cyclin-dependent kinase (*CDK*) complexes are targets for CDK inhibitors including p21[WAF1/CIP1] (*p21*) and p27[KIP1] (*p27*). Also shown schematically are approximate times of sensitivity to progestin stimulation and inhibition

for DNA replication before full expression of a differentiated phenotype after growth arrest. This alternative could account for the coordination in magnitude of the dual effects of progestins in breast cancer cells.

5.4 Mechanisms of Cell Cycle Control

Recent advances in delineating the molecular mechanisms underlying the control of cell proliferation identify cyclins and their catalytic partners the cyclin-dependent kinases (CDKs) as regulators of key transitions during the cell cycle (Hunter and Pines 1994; Sherr 1994). The sites of progestin action within G_1 phase suggest that regulation of the activity of cyclin D1/Cdk4 and cyclin E/Cdk2, which are major G_1 phase kinases in breast cancer cells, is a likely consequence, and possible cause, of alterations in the rate of cell proliferation following progestin treatment. Progress through G_1 phase is accompanied by sequential activation of these kinases and consequent phosphorylation of specific substrates: cyclin D1/Cdk4 is activated in mid-G_1 phase, while cyclin E/Cdk2 is activated toward the end of G_1 phase (Fig. 4). The physiological substrates for the G_1 phase CDKs have not been well defined but include the tumour suppressor protein pRB (Sherr 1994). This protein is present for much of G_1 in an underphosphorylated, growth inhibitory form but becomes hyperphosphorylated in late G_1, allowing entry into S phase.

As might be expected for enzymes integral to the control of a fundamental biological process, the activity of CDKs is subject to multiple levels of regulation (Morgan 1995). Cyclin binding is a prerequisite for activation. However, CDK activation also depends on the phosphorylation state of the kinase subunit, regulated by the activities of kinases and phosphatases including the CDK activating kinase (CAK). Finally, several low-molecular-weight proteins which bind cyclin/CDK complexes, acting as endogenous inhibitors of CDK activity, have been identified (Hunter 1993). One inhibitor, p27[KIP1], is postulated to provide a link between the sequential activation of Cdk4 and Cdk2, while another, p21[WAF1/CIP1], is present in all cyclin/CDK complexes (Fig. 4). Regulation of cyclin and CDK inhibitor abundance thus provides a mechanism for regulating CDK activity which is utilized not only to ensure orderly progression through the cell cycle but also to alter proliferation rate in response to external mitogenic or growth-inhibitory signals. Although CAK activity does not alter during progress through the cell cycle, regulation of the activity of a CAK (or one of the other kinases or phosphatases which target CDKs) is another possible means of regulating CDK activity.

5.5 Progestin Effects on CDK Function

Our first studies investigating the possible effects of progestins on CDK function focused on the stimulatory component of the response since we had previously defined the sequential induction of cyclin genes following growth factor stimulation of cell cycle progression in i
breast cancer cells (Musgrove et al. 1993). The entry of progestin-stimulated cells into S phase was preceded by an increase in cyclin D1 mRNA and protein abundance (Fig. 5). Cyclin D1 mRNA levels reached their peak after 6 h and thereafter declined, returning to control levels by 12 h (Fig. 5a). That the induction of cyclin D1 protein led to an increase in cyclin D1-associated kinase activity (predominantly due to cyclin D1/Cdk4 in T-47D cells) was confirmed by two experimental results. First, following the addition of progestin the proportion of pRB in the hyperphosphorylated form increased (Fig. 5b). Secondly, a more direct measurement of cyclin D1-associated kinase activity, using a recombinant pRB(379–928) fusion protein as a substrate, showed an increase in kinase activity after 6 h progestin treatment (Fig. 5b).

Addition of the progestin antagonist RU 486 either at the same time as, or 3 h subsequent to, progestin treatment rapidly reversed the increase in cyclin D1 mRNA abundance, although by 3 h a substantial increase had already occurred (Musgrove et al. 1993; Fig. 5a). This result is consistent with induction of cyclin D1 being a critical component of the mitogenic response to progestins, since the progestin antagonist addition is sufficient to prevent progestin-induced entry into S phase in either experimental design (Musgrove et al. 1991). This conclusion is strongly supported by experiments using T-47D breast cancer cells expressing cyclin D1 under the control of a inducible promoter, which show that cyclin D1 abundance is rate limiting for G_1 phase progression in these cells (Musgrove et al. 1994). Thus, a progestin-induced increase in cyclin D1 abundance can account for the acceleration in G_1 phase progression. It is important to note, however, that the induction of cyclin D1 is preceded by the induction of other genes (e.g. c-*myc*, Fig. 5a) and that the effect on cyclin D1 expression may not be direct, but rather result from prior induction of another gene or genes. Indeed, the 2 kb of the cyclin D1 gene promoter for which the sequence is available does not contain a classical progesterone response element.

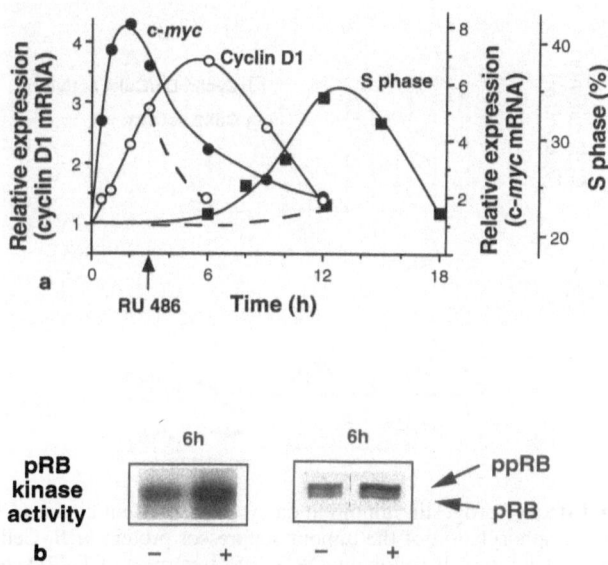

Fig. 5. a Effect of progestin stimulation on c-*myc* and cyclin D1 gene expression. T-47D cells were treated with ORG 2058 and harvested for northern blot analysis at intervals thereafter. Parallel flasks were treated with RU 486 which was added 3 h after progestin. The consequent effects on cyclin D1 and %S phase are shown in *dashed lines*. Redrawn from Musgrove et al. (1993).
b Effect of progestin stimulation on cyclin D1/Cdk4 activity and pRB phosphorylation. Cell lysates were harvested 6 h following ORG 2058 treatment of T-47D cells. The kinase activity of cyclin D1 immunoprecipitates was determined using a recombinant pRB(379–928) fusion protein as a substrate (*left*) and the hypophosphorylated and hyperphosphorylated forms of the tumour suppressor protein pRB (*pRB* and *ppRB*, respectively) distinguished by their different mobilities on sodium dodecyl sulphate-polyacrylamide gel electrophoresis followed by western blotting (*right*)

The effects of progestin-induced inhibition of cell proliferation on CDK function are less clear and are currently under investigation. Following more prolonged progestin treatment of T-47D cells, cyclin D1-associated kinase activity and Cdk2 activity began to decrease after 12 h and by 24 h had declined to 20 % of control (Fig. 6), a time course similar to that of the decrease in %S phase cells. These changes were associated with decreased phosphorylation of pRB. Until 18 h, although

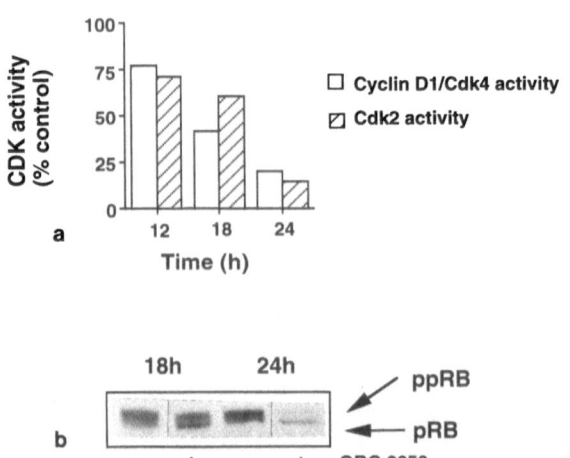

Fig. 6a,b. Effect of progestin inhibition on cyclin-dependent kinase (*CDK*) activity and phosphorylation of the tumour suppressor protein *pRB*. Cell lysates were harvested at intervals following ORG 2058 treatment of T-47D cells.
a The kinase activity of cyclin D1 and Cdk2 immunoprecipitates was determined using recombinant pRB(379–928) fusion protein and histone H1, respectively, as substrates. **b** Western blot of pRB, as in Fig. 5b

the proportion of hyperphosphorylated pRB decreased, both the hypo- and hyperphosphorylated forms were clearly present (Fig. 6). However, by 24 h essentially no hyperphosphorylated pRB was evident (Fig. 6), consistent with the low levels of both cyclin D1/Cdk4 and Cdk2 activity and the low %S phase associated with almost total inhibition of cell proliferation. The mechanisms underlying the decreased CDK activity remain to be defined. However, preliminary data suggest that changes in the abundance of cyclin D1, cyclin E, Cdk4 or Cdk2 are only minor within 24 h of progestin treatment and that other factors might contribute to the marked loss of CDK activity. One possibility is inhibition of CDK activity via induction of a CDK inhibitor, e.g. p21[WAF1/CIP1], which increases in abundance by approximately twofold following progestin treatment. Furthermore, while clear effects on CDK function are associated with progestin inhibition of proliferation, whether such changes are a cause or a consequence of the changes in the rate of cell proliferation remains to be defined.

5.6 Conclusions

Figure 7 presents a model for progestin actions on cell cycle progression, which was developed following our initial demonstration of the biphasic effect of progestin on cell cycle progression in T-47D cells (Clarke and Sutherland 1990). Progestin stimulation is proposed to result from the actions of a gene-denoted "start", the product of which controls the rate of G_1 phase progression. This gene may be regulated either directly or indirectly by progestins and, because of its critical role in regulating cell cycle progression, is likely to also be a target of other mitogens, including oestrogen and peptide growth factors. The multifactorial regulation of this gene favours an indirect means of regulation by progestins, via the induction of a "progesterone receptor-induced protein" (PRIP). Cyclin D1 is one of the very few proteins identified as rate limiting for G_1 phase progression in cells, including breast cancer cells (Musgrove et al. 1994; Sherr 1994) and cyclin D1 induction is an early response to mitogens, including oestrogens, progestins and a range

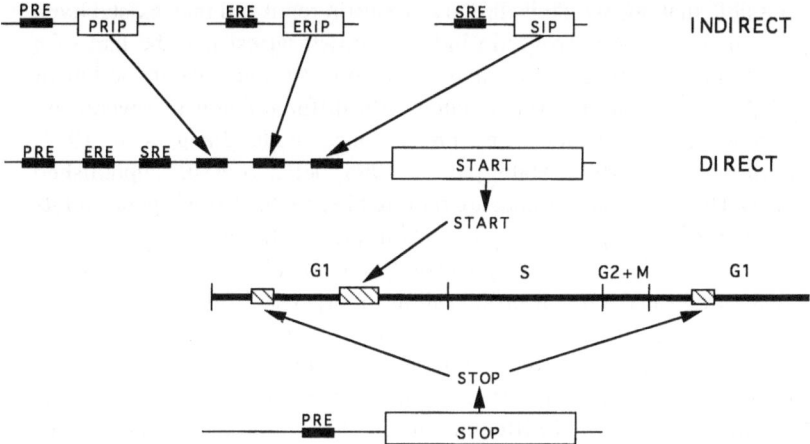

Fig. 7. Potential mechanisms for progestin regulation of cellular proliferation. *PRE*, progestin response element; *ERE*, oestrogen response element; *SRE*, serum response element; *PRIP*, *ERIP* and *SIP* are the products of genes induced by PRE, ERE and SRE, respectively; *M*, mitosis. See text for details. Redrawn from Clarke and Sutherland (1990)

of growth factors (Musgrove et al. 1993). Furthermore, there is a close correlation between regulation of cyclin D1 gene expression and subsequent acceleration into S phase following progestin treatment. Thus, cyclin D1 has the properties predicted for the "start" gene product.

Progestin inhibition is proposed to be mediated in a similar fashion, via induction of an inhibitory gene denoted "stop", perhaps as part of a differentiation programme. Again, this gene may be regulated either directly or indirectly. However, since very few gene products determine the rate of cell cycle progression but inhibition of any one of a wide range of processes can lead to growth arrest, "stop" is perhaps more likely than "start" to be directly regulated by progestins. Preliminary data suggest CDK inhibitors or regulators of CDK inhibitor expression and function as candidates for the stop gene. These molecules can be classified into two families based on their structure and function: one characterized by an ankyrin repeat structure and with specificity toward cyclin D-associated CDKs (e.g. $p16^{INK4}$ and $p15^{INK4B}$) and the other with inhibitory activity toward a broader range of CDKs (e.g. $p21^{WAF1/CIP1}$ and $p27^{KIP1}$) (Hunter 1993; Hunter and Pines 1994). Further CDK inhibitors may well remain to be identified and it is possible that one which is directly progestin regulated may exist. Given the suggestion that progestin-induced growth arrest may be part of a differentiation programme, it is interesting to note that induction of $p21^{WAF1/CIP1}$ has been associated with differentiation in several experimental models, including breast cancer cells (Jiang et al. 1994; Steinman et al. 1994; Halevy et al. 1995, deFazio et al., unpublished data). However, further data are required before the "stop" gene and its mechanisms of regulation by progestins are identified.

In conclusion, the data presented here identify the complexity of progestin effects on cell proliferation in target cells and provide some insight into potential mechanisms based on ongoing studies in one experimental model system. Together, these data provide a mechanism for the known inhibitory effects of progestins on cell proliferation, i.e. G_1 arrest and terminal differentiation, although the molecular basis of these effects remain to be fully defined. They also provide evidence for stimulation of cell cycle progression by progestins. Although both processes are mediated via the progesterone receptor and occur together in breast cancer cells, the fact that different genes are likely to be involved raises the possibility of cell-specific regulation of these genes resulting in stimulatory effects in some cell types and inhibitory effects on others. Further insight into these mechanisms should provide a deeper understanding of progestin regulation of cell proliferation and the likely consequences of prolonged progestin administration.

Acknowledgements. This research has been supported by grants from the National Health and Medical Research Council of Australia and the New South Wales State Cancer Council.

References

Borellini F, Oka T (1989) Growth control and differentiation in mammary epithelial cells. Environ Health Perspect 80:85–99

Clarke CL, Sutherland RL (1990) Progestin regulation of cellular proliferation. Endocr Rev 11:266–302

Clarke CL, Sutherland RL (1993) Progestin regulation of cellular proliferation: update 1993. In: Horwitz KB (ed) Endocrine aspects of cancer. Endocrine Society, Bethesda, pp 132–135

Creasy GW, Kafrissen ME, Upmalis D (1992) Review of the endometrial effects of estrogens and progestins. Obstet Gynaecol Surv 47:654–678

Das RM, Martin L (1973) Progesterone inhibition of mouse uterine epithelial proliferation. J Endocrinol 59:205–206

Gompel A, Malet C, Spritzer P, Lalardrie JP, Kuttenn F, Mauvais Jarvis P (1986) Progestin effect on cell proliferation and 17 β-hydroxysteroid dehydrogenase activity in normal human breast cells in culture. J Clin Endocrinol Metab 63:1174–1180

Halevy O, Novitch BG, Spicer DB, Skapek SX, Rhee J, Hannon GJ, Beach D, Lassar AB (1995) Correlation of terminal cell cycle arrest of skeletal muscle with induction of p21 by MyoD. Science 267:1018–1021

Hall RE, Lee CSL, Alexander IE, Shine J, Clarke CL, Sutherland RL (1990) Steroid hormone receptor gene expression in human breast cancer cells: inverse relationship between estrogen and glucocorticoid receptor messenger RNA levels. Int J Cancer 46:1081–1087

Hall RE, Tilley WD, McPhaul MJ, Sutherland RL (1992) Regulation of androgen receptor gene expression by steroids and retinoic acid in human breast-cancer cells. Int J Cancer 52:778–784

Haslam SZ (1988) Progesterone effects on deoxyribonucleic acid synthesis in normal mouse mammary glands. Endocrinology 122:464–470

Henderson BE, Ross RK, Pike MC (1991) Toward the primary prevention of cancer. Science 254:1131–1138

Horwitz KB, Freidenberg GR (1985) Growth inhibition and increase of insulin receptors in antiestrogen-resistant T47D$_{co}$ human breast cancer cells by progestins: implications for endocrine therapies. Cancer Res 45:167–173

Hulka BS, Liu E, Lininger RA (1994) Steroid hormones and the risk of breast cancer. Cancer 74:1111–1124

Hunter T (1993) Braking the cycle. Cell 75:839–841

Hunter T, Pines J (1994) Cyclins and cancer II: cyclin D and CDK inhibitors come of age. Cell 79:573–582

Jiang H, Lin J, Su Z-Z, Collart FR, Huberman E, Fisher PB (1994) Induction of differentiation in human promyelocytic HL-60 leukemia cells activates p21, WAF1/CIP1, expression in the absence of p53. Oncogene 9:3397–3406

Laidlaw IJ, Clarke RB, Howell A, Owen AWMC, Potten CS, Anderson E (1995) The proliferation of normal human breast tissue implanted into athymic nude mice is stimulated by oestrogen but not progesterone. Endocrinology 136:164–171

Martin L, Finn CA (1971) Oestrogen-gestagen interactions on mitosis in target tissues. In: Hubinot PO, Leroy F, Galard P (eds) Basic actions of sex steroids on target organs. Karger, Basel, pp 172–188

Martin L, Das RM, Finn CA (1973) The inhibition by progesterone of uterine epithelial proliferation in the mouse. J Endocrinol 57:549–554

Morgan DO (1995) Principles of CDK regulation. Nature 374:131–134

Musgrove EA, Sutherland RL (1993) Effects of the progestin antagonist RU 486 on T-47D cell cycle kinetics and cell cycle regulatory genes. Biochem Biophys Res Commun 195:1184–1190

Musgrove EA, Lee CSL, Sutherland RL (1991) Progestins both stimulate and inhibit breast cancer cell cycle progression while increasing expression of transforming growth factor α, epidermal growth factor receptor, c-fos and c-myc genes. Mol Cell Biol 11:5032–5043

Musgrove EA, Hamilton JA, Lee CSL, Sweeney KJE, Watts CKW, Sutherland RL (1993) Growth factor, steroid and steroid antagonist regulation of cyclin gene expression associated with changes in T-47D human breast cancer cell cycle progression. Mol Cell Biol 13:3577–3587

Musgrove EA, Lee CSL, Buckley MF, Sutherland RL (1994) Cyclin D1 induction in breast cancer cells shortens G$_1$ and is sufficient for cells arrested in G$_1$ to complete the cell cycle. Proc Natl Acad Sci USA 91:8022–8026

Russo IH, Russo J (1991) Progestagens and mammary gland development: differentiation versus carcinogenesis. Acta Endocrinol 125:7–12

Russo J, Russo IH (1993) Development pattern of human breast and susceptibility to carcinogenesis. Eur J Cancer Prev 2 [Suppl 3]:85–100

Sedlacek S, Horwitz K (1984) The role of progestins and progesterone receptors in the treatment of breast cancer. Steroids 45:467–484

Sherr CJ (1994) G1 phase progression: cycling on cue. Cell 79:551–555

Steinman RA, Hoffman B, Iro A, Guillouf C, Lieberman DA, El-Houseini ME (1994) Induction of p21 (WAF-1/CIP1) during differentiation. Oncogene 9:3389–3396

Sutherland RL, Hall RE, Pang GYN, Musgrove EA, Clarke CL (1988) Effect of medroxyprogesterone acetate on proliferation and cell cycle kinetics of human mammary carcinoma cells. Cancer Res 48:5084–5091

Vignon F, Bardon S, Chalbos D, Rochefort H (1983) Antiestrogenic effect of R5020, a synthetic progestin in human breast cancer cells in culture. J Clin Endocrinol Metab 56:1124–1130

Whitehead MI, Townsend PT, Pryse-Davies J, Ryder TA, King RJ (1981) Effects of estrogens and progestins on the biochemistry and morphology of the postmenopausal endometrium. N Engl J Med 305:1599–1605

6 Central Versus Endometrial Effects of Antiprogestins: Is Endometrial Selectivity Possible?

M. Bygdeman, K. Gemzell-Danielsson, and M. L. Swahn

6.1   Introduction 107
6.2   Follicular Development and Ovulation 108
6.2.1   Effect of Antiprogestin 108
6.2.2   Mode of Action of Antiprogestin 109
6.3   Endometrial Development and Function 110
6.3.1   Effect of Antiprogestin on Bleeding Patterns 110
6.3.2   Effect of Antiprogestin on Endometrial Maturation 111
6.4   Possible Difference in the Dose Inhibiting Ovulation
        and Endometrial Development 113
6.4.1   Contraceptive Use of Antiprogestin 113
6.4.2   Sensitivity in Follicular and Endometrial Development 114
6.5   Conclusions 116
References 116

6.1 Introduction

Progesterone is a key hormone in the regulation of many reproductive processes, including the establishment and maintenance of pregnancy. During the luteal phase, progesterone stimulates the development and maturation of the endometrium. It has long been recognized that compounds which interfere with synthesis of progesterone or block its action may offer a significant advantage in fertility control (Baulieu 1975), since exact synchrony between embryonic development and the

state of the endometrium seems to be essential for successful implantation (Navot et al. 1991). Several antiprogestins which bind to the progesterone receptor, thereby preventing progesterone from expressing its
biological effect, have been developed. Among these are mifepristone
(Roussel Uclaf, Paris, France) and onapristone and lilopristone (Schering AG, Berlin, Germany). The effect of these two compounds has
mainly been tested in various animal species, including nonhuman
primates. Mifepristone in humans and onapristone in bonnet monkeys
have both been shown to prevent pregnancy by affecting endometrial
development (Glasier et al. 1992; Gemzell-Danielsson et al. 1993; Katkam et al. 1995).

6.2 Follicular Development and Ovulation

6.2.1 Effect of Antiprogestin

It is well known that the administration of antiprogestin in the mid- or
late follicular phase delays the luteinizing hormone (LH) surge and
postpones ovulation. Estrogen levels fail to increase and follicular development is delayed or arrested. No bleeding is induced and endometrial development does not seem to be influenced (Liu et al. 1987;
Shoupe et al. 1987a; Luukainen et al. 1988; Swahn et al. 1988; Puri et
al. 1989). After cessation of mifepristone treatment, there is a resump-
tion of follicular growth or recruitment of new follicles, and the following luteal phase is usually normal (Swahn et al. 1988). Ovulation can be
postponed for several months if the treatment is continued (Yen 1993).

A daily dose of 2 mg is the minimum dose of mifepristone that seems
to suppress ovulation consistently, whereas 1 mg per day results in
various effects on cycle length. In some women, ovulation apparently
occurred normally; in others, ovulation was suppressed or evidence of
luteinized unruptured follicle was found (Ledger et al. 1992; Croxatto et
al. 1993). Once-weekly administration of 10 mg mifepristone still inter-
feres with normal follicular development and function (Spitz et al.
1993). We have recently shown that 5 mg or 2.5 mg mifepristone once
weekly for 2 months starting on cycle day 2 does not inhibit ovulation.
However, the length of the follicular phase could occasionally be prolonged for up to 13 days, whereby the length of the luteal phase and the

levels of ovarian steroids during this period was normal (Gemzell-Da-nielsson et al. 1995).

Similar results have been reported for onapristone. This antiprogestin was given to bonnet monkeys in doses ranging from 5 to 20 mg once weekly for two menstrual cycles. The lowest dose did not impair the length of the menstrual cycle or affect follicular development and ovulation. Plasma levels of progesterone and estradiol were unchanged (Ishwad et al. 1993).

6.2.2 Mode of Action of Antiprogestin

Antiprogestin may affect folliculogenesis and ovulation either directly on the ovary or indirectly through an effect on the hypothalamic pituitary axis, or both. In the human, there is strong evidence that the rising levels of estradiol from the dominant follicle is the main stimulus for the LH surge. In spite of this fact, administration of 1 mg mifepristone daily in late follicular phase delayed ovulation although the secretion of estradiol exceeded that normally present during a control cycle (Batista et al. 1992a). The inhibitory effect of mifepristone could be overcome by the simultaneous administration of progesterone (Batista et al. 1992b). In women with hypothalamic amenorrhea, ovulation was induced with gonadotropin-releasing hormone. If 1 mg mifepristone was given daily to these women for 5 days, when the follicular diameter was between 14 and 16 mm, the midcycle gonadotropin surge and ovulation were delayed (Batista et al. 1994). These findings indicate that the preovulatory increase in progesterone of ovarian origin might reinforce the positive feedback effect of estrogen in the triggering of the midcycle LH surge and that the inhibitory effect of mifepristone on ovulation is at least partly due to a blocking effect of progesterone at the pituitary level. The finding in bonnet monkeys that the inhibitory effect of onapristone on follicular development and ovulation could be overcome by treatment with LH and follicle-stimulating hormone (FSH) or human FSH alone supports this conclusion (Puri and Van Look 1991).

6.3 Endometrial Development and Function

6.3.1 Effect of Antiprogestin on Bleeding Patterns

As mentioned previously, the administration of antiprogestin during the follicular phase of the menstrual cycle does not seem to have any effect on the endometrium. The effect of antiprogestins during the luteal phase depends on the time of treatment and the dose of antiprogestin. Even if a high dose of mifepristone, i.e., 200 mg, is administered immediately after ovulation (day LH+2), no bleeding except for occasional spotting after 2–3 days is induced, and the length of the luteal phase and the levels of ovarian steroids are not significantly influenced (Swahn et al. 1990). Administration of high doses of mifepristone later during the luteal phase results in shedding of the endometrium and subsequent bleeding a few days after the initiation of treatment (Swahn et al. 1988; Li et al. 1988). The fact that the antiprogestin-induced bleeding is not associated with a decrease in plasma progesterone levels and can be induced even if corpus luteum function is supported by simultaneous administration of human chorionic gonadotrophin (hCG), strongly indicates that mifepristone directly affects the endometrium (Croxatto et al. 1985; Swahn et al. 1990). Since no bleeding occurs if mifepristone is administered during the follicular phase or immediately after ovulation, or if given to anovulatory women (Schaison et al. 1985; Swahn et al. 1988; Puri and Van Look 1991), progesterone influence and secretory transformation of the endometrium seem to be prerequisites for mifepristone-induced bleeding. The dose of mifepristone is also of importance since once-weekly administration of 5 mg mifepristone starting on cycle day 2 or 10 mg given on cycle days LH+5 and LH+8 did not induce any bleeding in spite of an inhibition of endometrial development (Gemzell-Danielsson et al. 1995; Greene et al. 1992). Similar findings have been reported for onapristone in the bonnet monkey (Puri et al. 1989).

The mechanism through which antiprogestins cause endometrial bleeding is not exactly known but it may involve alterations in endogenous prostaglandin production and inhibition of prostaglandin catabolism (Kelly et al. 1986) and endometrial aromatase activity (Tseng et al. 1986) resulting in degenerative changes in the vascular structure (Johannisson et al. 1989) and subsequent shedding of the endometrium.

In most subjects the mifepristone-induced bleeding is followed by a second bleeding around the expected time of menstruation which is preceded by a decrease in plasma progesterone levels (Li et al. 1988; Swahn et al. 1988). However, some women treated with mifepristone experience only one bleeding episode, which is associated with premature luteal regression (Shoupe et al. 1987b; Li et al. 1988; Swahn et al. 1988). The factors that determine whether luteolysis takes place are not well understood. There were no differences in dose of mifepristone given or resulting serum levels between subjects with one or two bleeding episodes (Shoupe et al. 1987b). The stage of the luteal phase when mifepristone is given seems to be of some importance (Swahn et al. 1988). However, most of the evidence available suggests that mifepristone-induced luteolysis results from inhibitory effects of hypothalamic and pituitary function as evidenced by a reduction in amplitude and frequency of LH pulses and blunting of the pituitary LH response to gonadotropin-releasing hormone (Schaison et al. 1985; Garzo et al. 1988; Shoupe et al. 1990).

6.3.2 Effect of Antiprogestin on Endometrial Maturation

Administration of antiprogestin early in the luteal phase has significant effects on endometrial development and function. The development and maturation of the endometrium is significantly inhibited, according to morphometric evaluation. A significantly decreased glandular diameter and an increased number of glandular and stromal mitoses have been demonstrated (Li et al. 1988; Swahn et al. 1990).

The concentration and distribution of estrogen and progesterone receptors undergo characteristic variations throughout the menstrual cycle. During the midfollicular period a small portion of glandular and stromal cells stains positively for progesterone receptors, while staining for estrogen receptors is more intense and frequent. During the late follicular and early luteal phase, glandular staining for progesterone receptors increases markedly and reaches its maximum. Thereafter, estrogen and progesterone receptor concentration decreases or disappears in glandular cells (Garcia et al. 1988; Lessey et al. 1988). The

Table 1. Some endometrial effects of antiprogestin in the luteal phase

Endometrial development	Delayed or inhibited
Volume of uterine secretion	Increased
Progesterone receptor concentration	Inhibition of downregulation
Concentration of $PGF_{2\alpha}$ in uterine secretion	Decreased
DBA lectin binding	Reduced
Placenta protein 14	Reduced

See text for references.

variations are believed to be due to the changes in plasma levels of estrogen and progesterone during the menstrual cycle and their different effects on receptor concentration, estrogen increasing and progesterone downregulating receptor concentration. Treatment with mifepristone significantly inhibits the normal downregulation of progesterone receptor concentration in the luteal phase (Table 1; Berthois et al. 1991; Mäentausta et al. 1993).

The effect of mifepristone on the secretory activity of the endometrium has been evaluated in different ways. The reduction in amount of uterine fluid at midluteal phase in comparison with that at ovulation time was less pronounced following treatment with mifepristone. Prostaglandins, either maternally or embryonically derived, have been suggested to be involved in the initial phase of implantation. Following mifepristone treatment, the concentration of prostaglandin F ($PGF_{2\alpha}$) in uterine fluid was significantly reduced (Gemzell-Danielsson and Hamberg 1994). The same is true for 17β-hydroxysteroid dehydrogenase, the major enzyme in the endometrium metabolizing estradiol into the biologically less active estrogen estrone (Mäentausta et al. 1993). The secretory components of the endometrium can be detected by lectin histochemistry using biotinylated *Dolichos biflorus* agglutinin (DBA). DBA binds to N-acetyl galactosamine and galactose residues present in the glandular secretion of the midluteal phase endometrium in a normal menstrual cycle (Mazur et al. 1981). Following treatment with 200 mg mifepristone on day LH+2 or 5 mg mifepristone once weekly, DBA–lectin staining in the luteal phase was weak or absent (Gemzell-Danielsson et al. 1994). The plasma concentration of placenta protein 14 (PP14) is another marker of endometrial function. Endometrial PP14 is synthesized by endometrial secretory glands. The serum PP14 levels rise during the last week of the luteal phase and peak at the onset of menstruation (Julkanen et al. 1986). After treatment with mifepristone the plasma concentration of PP14 was significantly reduced (Swahn et al. 1993; Gemzell-Danielsson et al. 1995).

6.4 Possible Difference in the Dose Inhibiting Ovulation and Endometrial Development

6.4.1 Contraceptive Use of Antiprogestin

As described previously, the two main effects of antiprogestin with regard to fertility control is inhibition of ovulation and endometrial development and function. Croxatto and Salvatierra (1991) have shown that with intermittent sequential therapy of mifepristone and a gestagen, it is possible to inhibit ovulation and maintain regular withdrawal bleeding.

A few clinical studies have shown that the effect of antiprogestin on endometrial development and function are effective in preventing implantation and pregnancy. In one study, 21 women were treated with 200 mg mifepristone on day LH+2 for up to 12 months (Gemzell-Danielsson et al. 1993). Overall, 169 cycles were studied, of which 157 were found to be ovulatory according to plasma progesterone concentration. On the basis of the LH peak, it was retrospectively calculated that in 124 cycles, at least one act of intercourse occurred during the period 3 days before to 1 day after ovulation with only one pregnancy occurring. Thus, the probability of pregnancy during this period of the menstrual cycle was 0.008, which is significantly lower than that reported in a WHO study (1983) in fertile women not using contraception.

Postcoital use of mifepristone has also been shown to be highly effective (Glasier et al. 1992; Webb et al. 1992). When 600 mg of mifepristone was administered within 72 h of one incidence of unprotected intercourse, no pregnancy occurred in the almost 600 women treated. There were significantly fewer side effects in terms of nausea and vomiting than with the conventional method, a combination of norgestrel and ethinyl estradiol. Based on the number of days that had elapsed since the last menstruation, a substantial number of women were treated at ovulation when the contraceptive effect of mifepristone must have been due to its inhibitory action on endometrial development and function. Both procedures have their disadvantages. Early luteal phase treatment permits repeated monthly administration; however, the main disadvantage is that the timing of the treatment is crucial. Although the LH surge can be self-measured by the woman, it may be regarded as too complicated or inconvenient. The cost of the LH test is also a limiting factor. Mifepristone can only be used occasionally after intercourse since it will upset the bleeding pattern of the menstrual cycle otherwise. If the treatment is given prior to ovulation, ovulation will only be delayed and it is necessary to use another contraceptive method or avoid additional intercourse during the remaining part of the cycle.

6.4.2 Sensitivity in Follicular and Endometrial Development

The results of these three studies (Gemzell-Danielsson et al. 1993; Webb et al. 1992; Glasier et al. 1992) strongly suggest that the endometrial effect of mifepristone is sufficient to prevent implantation. However, the doses of mifepristone (200–600 mg) are also sufficient to prevent ovulation. If the dose inhibiting endometrial function was lower than that influencing ovulation, one could imagine a contraceptive treatment based on very low doses of antiprogestin which do not affect ovarian function and the menstrual cycle. Since mifepristone mainly seems to influence ovarian function and ovulation by affecting the pituitary gland, while more directly affecting the endometrium through the withdrawal of the effect of progesterone , there may be a difference. Studies in the bonnet monkey have also demonstrated that once-weekly administration of 5 or 10 mg onapristone for two menstrual cycles did not inhibit ovulation but caused atrophic changes in the endometrial glands and stroma. The glands became smaller and inactive with apparent increase in the stroma. The pattern of midcycle rise in serum estradiol levels and progesterone levels during the luteal phase of both treatment cycles were comparable to those of vehicle-treated animals (Ishwad et al. 1993). A recent study from the same group of investigators with the same antiprogestin in bonnet monkeys indicates that this effect on the endometrium is sufficient to prevent implantation (Katkam et al. 1995).

A few studies indicate that a similar difference in ovarian and endometrial sensitivity to antiprogestin could also be present in the human. Treatment once weekly with 5 mg or 2.5 mg mifepristone for 2 months starting on cycle day 2 did not inhibit, but rather delayed ovulation in four out of 14 subjects by 6–13 days. The length of the luteal phase and the levels of ovarian steroids during this period of the menstrual cycle did not differ from those in the control cycle. A delay in endometrial development with a significant decrease in number of glands and glandular diameter and an inhibition of the downregulation of progesterone receptor concentration in the luteal phase of the cycle could be demonstrated. A significant reduction in DBA–lectin binding and in serum concentration of PP14, indicating inhibition of the normal secretory function of the endometrium, was also found (Gemzell-Danielsson et al. 1995).

In women treated with 1 mg mifepristone daily for 30 days starting on cycle day 1 or 2, the duration of the follicular phase was increased in two out of five cycles. However, suppressed ovulation during treatment with this dose was observed in only one of the five cycles studied. Endometrial maturation was disturbed in all cycles: one was inactive, one was proliferative, and three were asynchronous secretory (Croxatto et al. 1993). Batista et al. (1992a) used the same dose of mifepristone which was administered during days 1–25 of the menstrual cycle. In one out of ten cycles ovulation was suppressed while in the others the length of the follicular phase was delayed by 1–11 days (mean 6 days). In contrast, mifepristone did not change the duration of the luteal phase. Six of the nine women included in the data analyses had delayed endometrial development of 3–4 days during mifepristone treatment as opposed to none during placebo administration. The PP14 concentration around the onset of menstruation was significantly lower after mifepristone than after placebo treatment.

The results of these studies show that the endometrium is more sensitive to mifepristone than is the ovulatory process. This is especially true for the secretory activity of the endometrium. Not only endometrial development but also the secretory activity of the luminal and glandular epithelium may play an important role in implantation. Cyclic changes in the glycoprotein cell surface coat (glycocalyx) result in decreased electronegativity at the receptor stage due to replacement of highly acidic sulfated glycoprotein with moderately acidic sialyated glycoprotein (Jansen et al. 1985). Furthermore, production of an oligosaccharide epitope identified with the D9B1 monoclonal antibody is mildly diminished in the endometrial tissue of women with luteal phase defects (Seif et al. 1989). This is consistent with evidence that endometrial abnormalities in patients with unexplained infertility are due to impaired secretory activity of the endometrial glands and not related to the stromal components (Li et al. 1990; Klentzeris et al. 1991).

6.5 Conclusions

Antiprogestins may inhibit both ovulation and endometrial development and secretory function, depending on dose and time of treatment. The effect on ovulation may, at least partly, be due to an effect on pituitary release of gonadotropins while the effect on the endometrium is mainly a direct one. The inhibition of endometrial development is sufficient to prevent implantation. According to results from studies in subhuman primates and humans, it seems possible to develop dose schedules which inhibit endometrial development, and especially endometrial secretion, but not ovulation. Whether these effects on the endometrium are sufficient to prevent implantation remains to be established.

Acknowledgements. These studies were supported by the Knut and Alice Wallenberg Foundation and by the Swedish Medical Research Council (Project 05696)

References

Batista MC, Cartledge TP, Zellmer AW, Merino MJ, Axiotis C, Loriaux DL, Nieman LK (1992a) Delayed endometrial maturation induced by daily administration of the antiprogestin RU 486. A potential new contraceptive strategy. Am J Obstet Gynecol 167:60–65

Batista MC, Cartledge TP, Zellmer AW, Nieman LK, Merriam GR, Loriaux DL (1992b) Evidence for a critical role of progesterone in the regulation of the midcycle gonadotropin surge and ovulation. J Clin Endocrinol Metab 74:565–570

Batista MC, Cartledge TP, Zellmer AW, Nieman LK, Loriaux DL, Merriam GR (1994) The antiprogesterone RU 486 delays the midcycle gonadotrophin surge and ovulation in gonadotrophin-releasing hormone induced cycles. Fertil Steril 62:28–34

Baulieu EE (1975) Antiprogesterone effect and midcycle (periovulatory) contraception. Eur J Obstet Gynecol Reprod Biol 4/5:161–166

Berthois Y, Salat-Baroux J, Cornat J, De Brux J, Kopps F, Maric Martin P (1991) A multiparametric analysis of endometrial estrogen and progesterone receptors after postovulatory administration of mifepristone. Fertil Steril 55:547–554

Croxatto HB, Spitz JM, Salvatierra AM, Bardin CW (1985) The demonstration of the antiprogestin effects of RU 486 administered to the human during hCG-induced pseudopregnancy. In: Baulieu EE, Segal SJ (eds) The antiprogestin steroid RU 4886 and human fertility control. Plenum, New York, pp 263–271

Croxatto HB, Salvatierra AM (1991) Cyclic use of antigestagen for fertility control. In: Runnebaum B, Rabe T, Kiesel L (eds) Female contraception and male fertility regulation. Parthenon, London, pp 145–152

Croxatto HB, Salvatierra AM, Croxatto HD, Fuentealba B (1993) Effects of continuous treatment with low dose mifepristone throughout one menstrual cycle. Hum Reprod 8:201–207

Garcia E, Bouchard P, De Brus J, Berdah J, Frydman R, Schaison G Milgrom E, Perrot-Applanat M (1988) Use of immunocytochemistry of progesterone and estrogen receptors for endometrial dating. J Clin Endocrinol Metab 67:80–87

Garzo VG, Liu J, Ulmann A, Baulieu .E, Yen SSC (1988) Effects of an antiprogesterone (RU 486) on the hypothalamic-hypophyseal-ovarian-endometrial axis during the luteal phase of the menstrual cycle. J Clin Endocrinol Metab 66:506–517

Gemzell-Danielsson K, Hamberg M (1994) The effect of antiprogestin (RU 486) and prostaglandin biosynthesis inhibitor (Naproxen) on uterine fluid $PGF_{2\alpha}$ concentrations. Hum Reprod 9:1626–1630

Gemzell-Danielsson K, Swahn ML, Svalander P, Bygdeman M (1993) Early luteal phase treatment with RU 486 for fertility regulation. Hum Reprod 8:870–873

Gemzell-Danielsson K, Svalander P, Swahn ML, Johannisson E, Bygdeman M (1994) Effect of a single postovulatory dose of RU 486 on the endometrial maturation in the implantation phase. Hum Reprod 9:2398–2404

Gemzell-Danielsson K, Westlund P, Johannisson E, Swahn ML, Seppälä M, Bygdeman M (1995) Effect of low weekly doses of mifepristone on ovarian function and endometrial development. Hum Reprod (submitted)

Glasier AF, Thong KJ, Dewar M, Mackie M, Baird DT (1992) Mifepristone (RU 486) compared with high-dose estrogen and progesterone emergency postcoital contraception. N Engl J Med 327:1041–1044

Greene KE, Kettel LM, Yen SSC (1992) Interruption of endometrial maturation without hormonal changes by an antiprogesterone during the first half of the luteal phase of the menstrual cycle: a contraceptive potential. Fertil Steril 58:338–343

Ishwad PC, Katkam RR, Hinduja IN, Chwalisz K, Elger W, Puri CP (1993) Treatment with a progesterone antagonist ZK 98.299 delays endometrial development without blocking ovulation in bonnet monkeys. Contraception 48:57–70

Jansen RPS, Turner M, Johannisson E, Landgren BM, Diczfalusy E (1985) Cyclic changes in human endometrial surface glycoproteins: a quantitative histochemical study. Fertil Steril 44:85–91

Johannisson E, Oberholzer M, Swahn ML, Bygdeman M (1989) Vascular changes in the human endometrium following the administration of the progesterone antagonist RU 486. Contraception 39:103–117

Julkanen M, Apter D, Seppälä M, Stenman UH, Bohn H (1986) Serum levels of placental protein 14 reflect ovulation in nonconceptional menstrual cycles. Fertil Steril 45:47–50

Katkam RR, Gopalkrishnan K, Chwalisz K, Schillinger E, Puri CP (1995) Onapristone (ZK 98,299). A potential antiprogestin for endometrial contraception. Am J Obstet Gynecol (in press)

Kelly RW, Healy DL, Cameron MJ, Cameron IT, Baird DT (1986) The stimulation of prostaglandin production by two antiprogesterone steroids in human endometrial cells. J Clin Endocrinol Metab 62:1116–1123

Klentzeris LD, Bulmer JN, Li TC, Morrison L, Warren A, Cooke ID (1991) Lectin binding of endometrium in women with unexplained infertility. Fertil Steril 56:660–667

Ledger WL, Sweeting VM, Hillier H, Baird DT (1992) Inhibition of ovulation by low dose mifepristone (RU 486). Hum Reprod 7:945–950

Lessey BA, Killam AP, Metager DA, Hanney AF, Greene GL, McCarty KS Jr (1988) Immunohistochemical analysis of human uterine estrogen and progesterone receptors throughout the menstrual cycles. Clin Endocrinol Metab 67:334–340

Li T-C, Dockery P, Thomas P, Rogers AW, Lenton EA, Cooke ID (1988) The effects of progesterone blockade in the luteal phase of normal fertile women. Fertil Steril 50:732–742

Li T-C, Dockery P, Rogers AW, Cooke ID (1990) A quantitative study of endometrial development in the luteal phase: comparison between women with unexplained infertility and normal fertility. Br J Obstet Gynaecol 97:576–582

Liu JH, Garzo G, Monis Strenkel C, Ulmann A, Yen SCC (1987) Disruption of follicular maturation and delay of ovulation after administration of the antiprogestin RU 486. J Clin Endocrinol Metab 65:1135–1140

Luukainen T, Heikinheimo O, Haukkamaa M, Lähteenmäki P (1988) Inhibition of folliculogenesis and ovulation by the antiprogesterone RU 486. Fertil Steril 49:361–363

Mazur MT, Duncan DA, Younger JB (1981) Endometrial biopsy in the cycle of conception: histologic and lectin histochemical evaluation. Fertil Steril 51:764–769

Mäentausta O, Svalander P, Gemzell-Danielsson K, Bygdeman M, Vikho R (1993) The effects of an antiprogesterone, mifepristone, and an antiestrogen, tamoxifen, on endometrial 17β-hydroxysteroid dehydrogenase and progestin and estrogen receptors during the luteal phase of the menstrual cycle. An immunohistochemical study. J Endocrinol Metab 77:913–918

Navot D, Scott TR, Droesch K, Veeck LL, Hung-Ching L, Rosenwaks Z (1991) The window of embryo transfer and the efficiency of human conception in vitro. Fertil Steril 55:114–118

Puri CP, Van Look PFA (1991) Newly developed competitive progesterone antagonists for fertility control. In: Agarwal MK (ed) Antihormones in health and disease. Karger, Basel, pp 127–167 (Frontiers of hormone research, vol 19)

Puri CP, Patel RK, Elger AG, Vadigoppula AD, Pongula JMR (1989) Gonadal and pituitary responses to progesterone antagonist ZK 98,299 during the follicular phase of the menstrual cycle in bonnet monkeys. Contraception 39:227–243

Schaison G, George M, Lestrat N, Reinberg A, Baulieu EE (1985) Effects of the antiprogesterone steroid RU 486 during midluteal phase in normal women. J Clin Endocrinol Metab 61:484–489

Seif M, Aplin JD, Buckely CH (1989) Luteal phase defect of an immunohistochemical diagnosis. Fertil Steril 51:273–279

Shoupe D, Mishell DR, Page MA, Madkour H, Spitz JM, Lobo RA (1987a) Effects of the antiprogesterone RU 486 in normal follicular phase. Am J Obstet Gynecol 157:1421–1426

Shoupe D, Mishell DR Jr, Lähteenmäki P, Heikinheimo O, Birgerson L, Madkour H, Spitz IM (1987b) Effects of the antiprogesterone RU 486 in normal women. I. Single-dose administration in the midluteal phase. Am J Obstet Gynecol 157:1415–1420

Shoupe D, Mishell DR Jr, Fossum G, Bopp BL, Spitz IM, Lobo RA (1990) Antiprogestin treatment decreases midluteal luteinizing hormone peak amplitude and primarily exerts a pituitary inhibition. Am J Obstet Gynecol 163:1982–1985

Spitz JM, Croxatto HB, Salvatierra AM, Heikinheimo O (1993) Response to intermittent RU 486 in women. Fertil Steril 59:571–575

Swahn ML, Johannisson E, Daniore V, de la Torre B, Bygdeman M (1988)The effect of RU 486 administered during the proliferative and secretory phase of the cycle on the bleeding pattern, hormonal parameters and the endometrium. Hum Reprod 3:915–921

Swahn ML, Bygdeman M, Xing S, Cekan S, Masironi B, Johannisson E (1990) The effect of RU 486 administered during the early luteal phase on bleeding pattern, hormonal parameters and endometrium. Hum Reprod 5:402–408

Swahn ML, Bygdeman M, Seppälä M, Johannisson E, Cekan S (1993) Effect of tamoxifen alone and in combination with RU 486 on the endometrium in the midluteal phase. Hum Reprod 8:93–200

Tseng L, Mazella J, Sun B (1986) Modulation of aromatase activity in human endometrial stromal cells by steroids, tamoxifen and RU 486. Endocrinology 118:1312–1318

Webb AMC, Russel J, Elstein M (1992) Comparison of Yuzpe regimen, danazol and mifepristone (RU 486) in oral postcoital contraception. Br Med J 305:927–931

WHO (1983) A prospective multicenter trial of the ovulation method of natural family planning. III. Characteristics of the menstrual cycle and of the fertile phase. Fertil Steril 40:773–778

Yen SCC (1993) Use of antiprogestins in the management of endometriosis and leiomyoma. In: Donaldson MS, Dorflinger L, Brown SB, Benel LZ (eds) Clinical applications of mifepristone and other antiprogestins. National Academy Press, Washington, pp 189–199

7 Androgen Action on the Bone*

J. S. Finkelstein

7.1   Introduction 121
7.2   Effects of Androgen Deficiency on Bone Mass in Men 122
7.2.1   Effects of Androgens in Normal Men 122
7.2.2   Bone Density in Hypogonadal Men 122
7.3   Areas for Future Investigation 132
7.3.1   Mechanism of Action of Androgens on Bone 132
7.3.2   Therapy of Androgen-Deficiency Bone Loss 134
References 135

7.1 Introduction

Osteoporosis is one of the leading causes of morbidity and mortality in the elderly. Osteoporosis affects 20 million Americans and leads to approximately 1.5 million fractures each year (Finkelstein 1995). The annual cost of health care and lost productivity attributed to osteoporosis exceeds $10 billion in the United States. During the course of their lifetimes, women lose about 50% of their trabecular bone and 30% of the cortical bone while men lose about 30% of their trabecular bone and 20% of their cortical bone. Thus, even though osteoporosis is less common in men than in women, one fifth of all hip fractures occur in men and by the age of 90 one of every six men will have fractured his hip.

---

* This chapter has been published previously in the Proceedings of the Second International Congress on Androgens and has been reproduced with the permission of John Wiley & Sons, Inc.

7.2 Effects of Androgen Deficiency on Bone Mass in Men

7.2.1 Effects of Androgens in Normal Men

In adults, bone density at any point in time is determined both by the peak bone mass achieved during development and the subsequent amount of bone loss. Androgens, by affecting both of these processes, are an important determinant of bone mass in men. Both cortical and trabecular bone density increase dramatically during puberty in boys (Krabbe and Christiansen 1984; Bonjour et al. 1991). The pubertal rise in testosterone is followed closely by an increase in serum alkaline phosphatase, a marker of osteoblast function, and subsequently bone density increases. These data strongly suggest that testosterone, or one of its metabolites, is responsible for the rise in bone mineral density during puberty. Peak trabecular bone density is usually achieved by the age of 18 years in males (Bonjour et al. 1991) though peak cortical bone density may not be reached for a few more years. Bone density then remains relatively stable in young adult males before it declines slowly in later life. Although several cross-sectional studies have suggested that a decline in gonadal function may be responsible for the decrease in bone density as men age, not all studies have been able to demonstrate a correlation between serum androgens and bone mass in aging men. Longitudinal studies of bone density in older adult men are needed to assess the relationship between changes in gonadal function and bone mass more precisely.

7.2.2 Bone Density in Hypogonadal Men

The observation that androgen deficiency could produce osteoporosis in men was first made by Albright, who noted that eunuchs often developed osteoporosis. With the advent of improved methods for measuring bone density, most of the data examining the relationship between androgen deficiency and bone mass in men have been published in the last 15 years. A series of reports involving a small number of patients suggested that cortical bone density is decreased in men with Klinefelter's syndrome. However, it is difficult to determine whether the osteopenia of these men is due to their hypogonadism or is an independent effect related to their genetic defect. In a case control study of 105 men with osteoporotic fractures, hypogonadism was one of the most common underlying disorders (Seeman et al. 1983). Another case control study reported that hypogonadism was twice as common in men with recent minimal trauma hip fractures than in controls. Histomorphometric analyses of iliac crest bone biopsies from men with androgen deficiency have shown both decreased and increased bone formation. This variability may be due to patient heterogeneity or to complex effects of androgens on bone turnover.

Comprehensive studies of the effects of androgen deficiency on bone mass in men have recently been reported in several groups of men: men with primary hypogonadism, men with several forms of secondary hypogonadism, and men with histories of constitutionally delayed puberty. These studies provide prospective data on the effects of androgens on bone mass in men and will be discussed in detail.

7.2.2.1 Studies in Men with Primary Hypogonadism

In contrast to women, for whom there is a large amount of data relating primary hypogonadism to bone mass, little such data exist for men. Stepan et al. (1989) measured bone mineral density of the lumbar spine using dual photon absorptiometry in 12 men who had undergone bilateral orchiectomy because of sexual delinquency up to 11 years previously (mean 5.6 years). In nine of these men, measurements were repeated 1–3 years later. As shown in Fig. 1, bone density decreased progressively with increasing number of years after castration. Although it appeared that the rate of bone loss was greater in the first several years after orchiectomy, the number of observations was too small to demonstrate such a relationship conclusively. Biochemical indices of bone resorption and bone formation were increased compared to normal men and there was a significant association between urinary hydroxyproline excretion and the rate of bone loss. These findings suggest that gonadal steroid deficiency is associated with increased bone turnover. All of the measured indices of bone turnover decreased during treatment with intranasal calcitonin, an agent that inhibits bone resorption.

Recently, we have investigated bone density and the effects of androgen replacement therapy in 29 men with acquired primary or secondary hypogonadism (Katznelson et al. 1994). Spinal bone density was

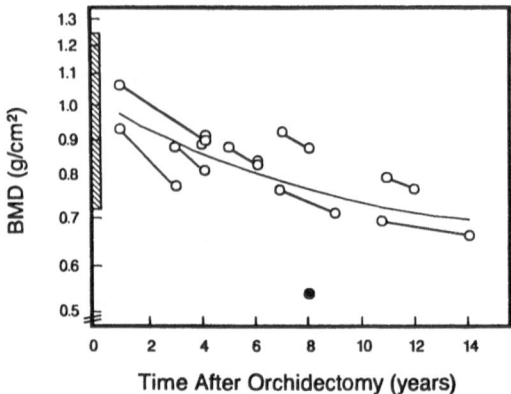

Fig. 1. Scattergram of lumbar spinal bone mineral density (*BMD*) as a function of time after orchidectomy in 12 men. In eight patients the measurement was repeated after 1–3 years. The *hatched bar* indicates the normal range for 20 men matched for age. The *solid circle* represents the value for one man who developed a hip fracture. Reproduced with permission from Stepan et al. (1989)

measured using both dual energy X-ray absorptiometry (DXA), a technique that assesses both trabecular and cortical bone in the spine, and quantitative computed tomography (QCT), a technique that assesses exclusively trabecular bone. Spinal bone density was significantly lower than that of age-matched normal men using both techniques. Spinal bone density increased significantly (5% by DXA and 13% by QCT) during 12–18 months of androgen replacement therapy. Baseline lean body mass was decreased in the hypogonadal men. Lean body mass increased and the percent body fat decreased in response to testosterone replacement.

7.2.2.2 Studies in Men with Hyperprolactinemic Hypogonadism
Greenspan et al. (1986) measured cortical bone density in the forearm by single photon absorptiometry (SPA) and trabecular bone density in the spine by QCT in 18 men between the ages of 30 and 79 with secondary hypogonadism caused by prolactin-secreting tumors. Five men had secondary hypothyroidism and/or secondary adrenal insufficiency and were receiving physiological hormone replacement. Both

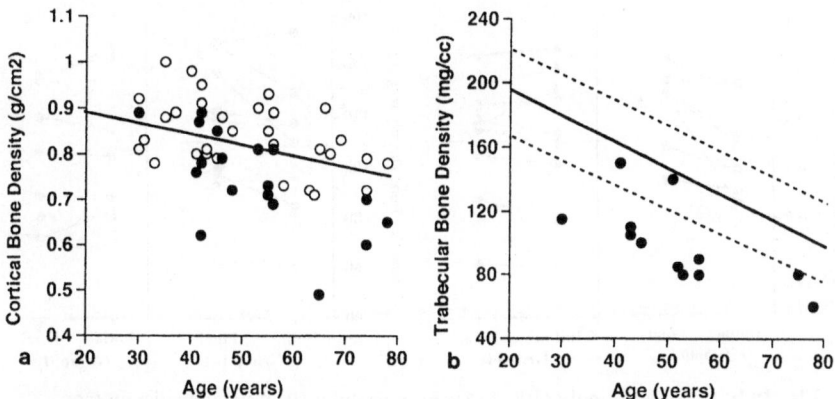

Fig. 2. a Cortical bone density of the radius in patients with hyperprolactinemia (*solid circles*) and controls (*open circles*) with the regression of cortical bone density with age in controls (*solid line*). **b** Trabecular bone density of the lumbar spine in patients with hyperprolactinemia (*solid circles*). The *solid line* represents the expected mean normal bone density for age (± SD). Modified and reproduced with permission from Greenspan et al. (1986)

cortical (Fig. 2a) and trabecular (Fig. 2b) bone mineral density were significantly decreased in the hyperprolactinemic men compared to age-matched controls. Cortical bone density correlated with the duration of hyperprolactinemia. There was no significant correlation between cortical and trabecular bone density, suggesting that cortical and trabecular bone respond differently to androgen deficiency. These findings suggest that hypogonadism in hyperprolactinemic men leads to osteopenia but do not rule out the possibility that other hormone deficiencies present in men with central hypogonadism may have an adverse effect on skeletal integrity.

To assess the effects of restoration of gonadal steroid secretion on bone density in men with hyperprolactinemic hypogonadism, Greenspan et al. (1989) performed serial measurements of bone density for 6–48 months in 20 such men who were treated with bromocriptine, transsphenoidal surgery, and/or cranial radiation. In those men whose serum testosterone levels normalized (group I), cortical bone density increased significantly (Fig. 3a) and there was a significant correlation between the change in serum testosterone levels and the change in bone

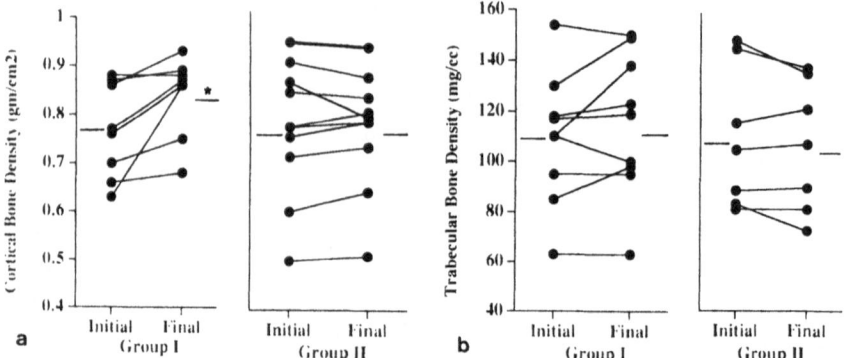

Fig. 3a,b. Initial and final cortical (**a**) and trabecular (**b**) bone densities in men with hyperprolactinemia in whom gonadal status normalized (group I) or who remained hypogonadal (group II). *Horizontal bars* represent the mean bone density. * $p<0.05$ vs initial value. Modified and reproduced with permission from Greenspan et al. (1989)

density. In the men who remained hypogonadal (group II), cortical bone density did not change (Fig. 3a). Trabecular bone density of the lumbar spine did not change significantly in either group (Fig. 3b).

7.2.2.3 Studies in Men Receiving Long-Acting Gonadotrophin Hormone-Releasing Hormone Analog Therapy

The effects of secondary hypogonadism on bone mass have also been studied by examining the effects of daily administration of a long-acting gonadotrophin hormone-releasing hormone (GnRH) analog to men with benign prostatic hyperplasia (Goldray et al. 1993). GnRH analog therapy produced severe testosterone deficiency in all men. In 10 of 17 men, bone density of the lumbar spine decreased significantly over a period of 6–12 months (Fig. 4). Serum alkaline phosphatase and osteocalcin levels increased significantly, suggesting that bone turnover was increased. Serum levels of calcium, vitamin D metabolites, and parathyroid hormone (PTH) did not change significantly. In general, the effects of GnRH analog administration on bone metabolism in men were similar to those previously described in women.

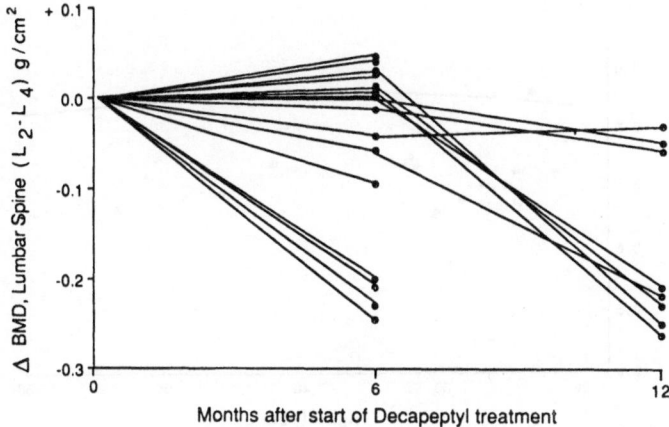

Fig. 4. Individual changes (Δ) in lumbar bone mineral density (*BMD*) during 6–12 months of gonadotrophin hormone-releasing hormone analog therapy. Reproduced with permission from Goldray et al. (1993)

7.2.2.4 Studies in Men with Idiopathic Hypogonadotropic Hypogonadism

Men with idiopathic hypogonadotropic hypogonadism (IHH) are hypogonadal due to an isolated absence of hypothalamic GnRH secretion. Otherwise, pituitary function is intact in these men. Thus, they provide a useful model to examine the effects of complete, isolated gonadal steroid deficiency on bone mass in men. Furthermore, because IHH is almost always a congenital abnormality, men with IHH provide a valuable model to assess the effects of gonadal steroid deficiency on pubertal bone development (i.e., the attainment of peak bone mass). We measured cortical bone density by SPA and trabecular bone density by QCT in 23 young men with IHH (Finkelstein et al. 1987). Because bone density increases dramatically during puberty, patients with open epiphyses were compared to adolescent controls matched for bone age, whereas patients with fused epiphyses were compared to age-matched adult men. Both cortical (Fig. 5) and trabecular (Fig. 6) bone mineral density were markedly decreased in IHH men and the osteoporosis was equally severe in the men with open and fused epiphyses. In eight men, trabecular bone density was below the fracture threshold despite their young age (Fig. 6). As in the men with hyperprolactinemic hypogonad-

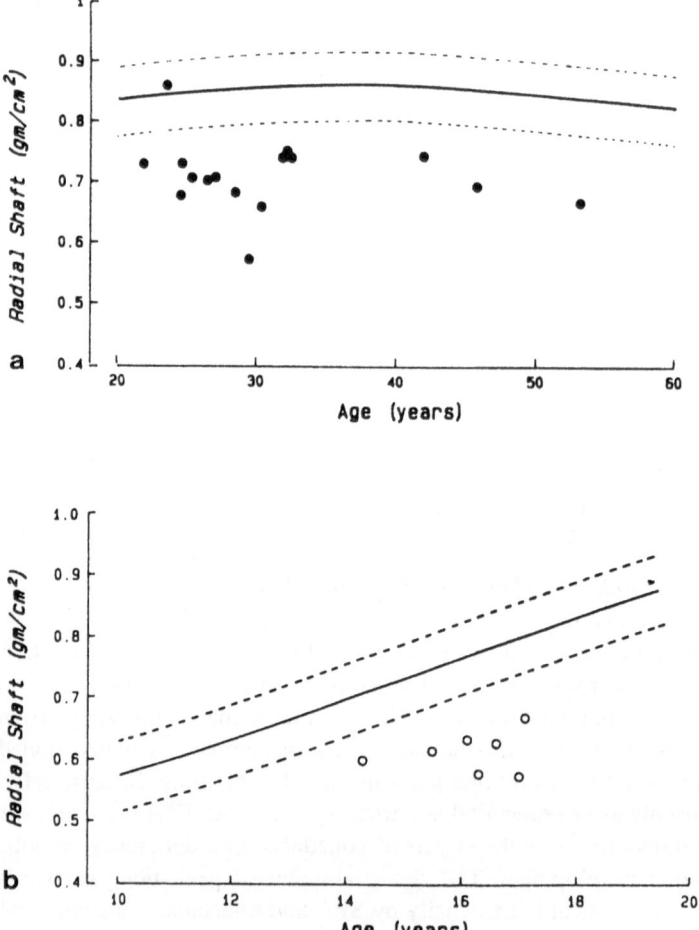

Fig. 5. a Radial (cortical) bone density compared with age in 16 men with idiopathic hypogonadotropic hypogonadism (IHH) with fused epiphyses. The *lines* indicate the mean (± 1 SD) radial bone density in normal men. **b** Radial bone density compared with bone age in seven men with IHH with open epiphyses. Modified and reproduced with permission from Finkelstein et al. (1987)

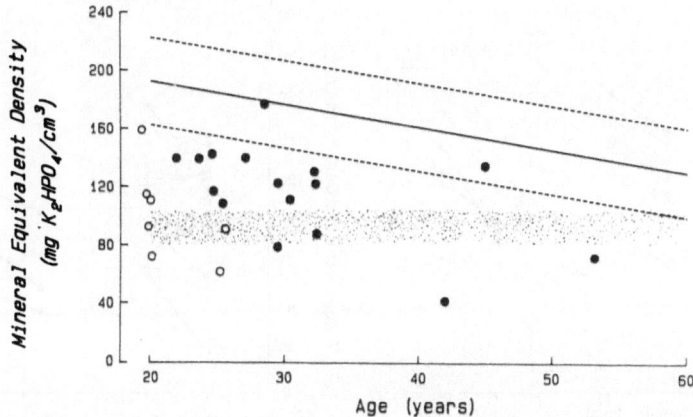

Fig. 6. Spinal (trabecular) bone density compared with age in 23 men with idiopathic hypogonadotropic hypogonadism (IHH). *Closed circles* refer to patients in group I (fused epiphyses), and *open circles* refer to patients in group II (open epiphyses). The *lines* indicate the mean (± 1 SD) spinal bone density in normal adult men. The *stippled bar* indicates the fracture threshold. Reproduced with permission from Finkelstein et al. (1987)

ism, there was no association between cortical and trabecular bone density. These data demonstrate that GnRH-deficient men have severe osteopenia affecting both cortical and trabecular bone. Because severe osteopenia was already present in men who were skeletally immature, these data suggest that the osteopenia of IHH men is due to inadequate pubertal bone accretion rather than postmaturity adult bone loss.

To assess the effects of androgen replacement on bone mass in IHH men, we made longitudinal measurements of cortical and trabecular bone density in 21 of these 23 men while serum testosterone levels were maintained in the normal range with either pulsatile GnRH, human chorionic gonadotropin, or intramuscular testosterone therapy for an average of 2 years (Finkelstein et al. 1989). In the men who were initially skeletally mature (group I), there was a small, but significant, increase in cortical bone density (Fig. 7a) whereas trabecular bone density did not change (Fig. 7b). These responses are similar to those described above for adult men with hyperprolactinemic hypogonadism. In the men who were still skeletally immature at the beginning of the

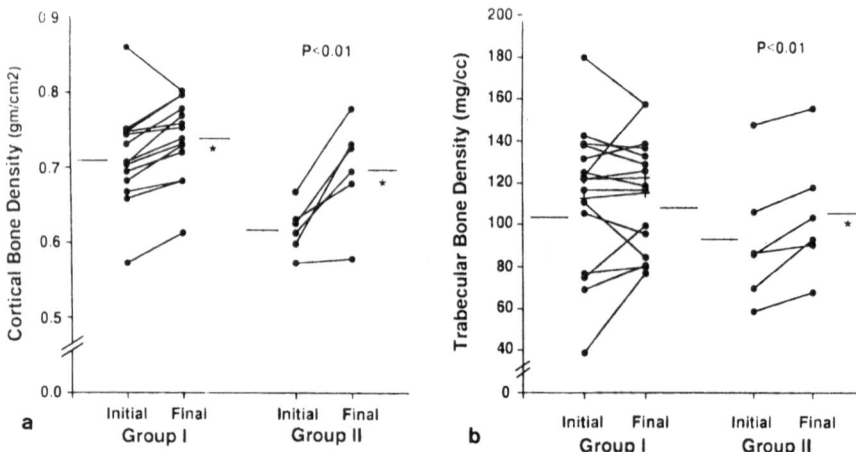

Fig. 7a,b. Initial and final cortical (**a**) and trabecular (**b**) bone densities after treatment of men with idiopathic hypogonadotropic hypogonadism (IHH) who initially had fused epiphyses (group I) or open epiphyses (group II). *Horizontal bars* represent the mean bone density. Modified and reproduced with permission from Finkelstein et al. (1989)

study (Group II), both cortical and trabecular bone density increased significantly (Fig. 7a,b). Furthermore, cortical bone density increased more in the skeletally immature men, suggesting that much of the increase in bone density in men with open epiphyses is due to a completion of the process of pubertal bone accretion. Despite these increases, bone density remained well below the levels of normal men. This finding suggests either that factors other than gonadal steroid deficiency are involved in the pathogenesis of the osteopenia of IHH men or that there may be a critical period in development when gonadal steroid secretion must be normal in order to achieve a normal peak bone density.

Fig. 8a,b. Radial bone density (**a**) and spinal bone density (**b**) in 23 men with ▶ histories of delayed puberty and 21 normal controls. The *horizontal lines* indicate the group means, and the *shaded areas* the mean ± 1 SD and ± 2 SD for the normal men. Reproduced with permission from Finkelstein et al. (1992)

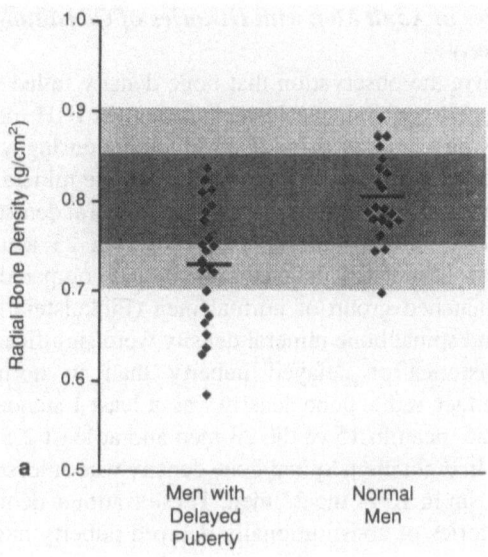

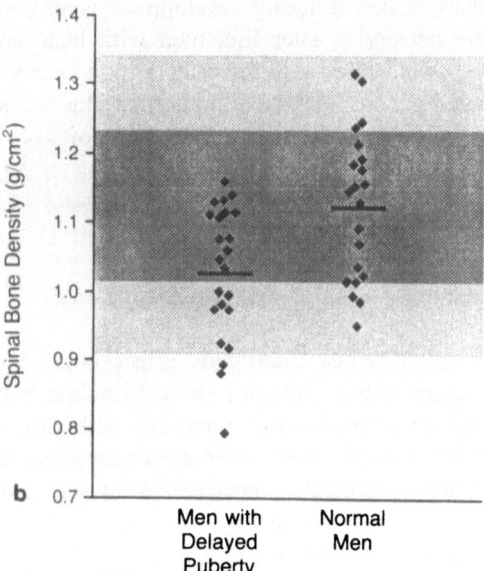

Fig. 8a,b. Legend see p. 130

7.2.2.5 Studies in Adult Men with Histories of Constitutionally Delayed Puberty

As noted above, the observation that bone density failed to normalize during prolonged gonadal steroid replacement in IHH men suggested that there may be a critical period in development during which puberty must occur in order to achieve a normal peak bone mineral density. To test this hypothesis, we measured radial bone mineral density using SPA and spinal bone mineral density using DXA in 23 adult men with histories of constitutionally delayed puberty and compared their values with a well-matched group of normal men (Finkelstein et al. 1992). Both radial and spinal bone mineral density were significantly lower in men with histories of delayed puberty than in normal controls (Fig. 8a,b). In fact, radial bone density was at least 1 standard deviation (SD) below the mean in 15 of the 23 men and at least 2 SD below the normal mean in eight men. Spinal bone density was at least 1 SD below the normal mean in 10 of the 23 men. These findings demonstrate that men with histories of constitutionally delayed puberty have decreased radial and spinal bone mineral density and suggest that the timing of puberty is an important determinant of peak bone mass. Because the peak bone density achieved during development is an important determinant of bone density in later life, men with histories of delayed puberty may be at increased risk for osteoporotic fractures. Finally, a history of delayed puberty may be an important clue to the etiology of low bone density in men with "idiopathic" osteoporosis.

7.3 Areas for Future Investigation

7.3.1 Mechanism of Action of Androgens on Bone

The mechanism(s) whereby androgens affect bone density is still unclear. Data suggest that androgens may stimulate bone formation directly. Several observations are consistent with this notion. First, androgen receptors have been found on osteoblasts. Second, both aromatizable and nonaromatizable androgens stimulate proliferation of human osteoblasts in vitro, as indicated by enhanced uptake of [^{3}H]thymidine. Androgens also stimulate collagen production in vitro. Third, androgens stimulate osteoblast differentiation in vitro, as indicated by an increase in the percentage of cells that stain for alkaline phosphatase (Kasperk et al. 1989). The effect of androgens on osteoblast proliferation and differentiation might be due to increased local production of transforming growth factor (TGF)-β or increased sensitivity to the effects of fibroblast growth factor and insulin-like growth factor (IGF)-II. Finally, androgens inhibit PTH-stimulated accumulation of cAMP by osteoblasts in vitro.

Androgens have also been shown to affect bone formation in humans, though it is not clear if the effects result from a direct action on osteoblasts or through an indirect mechanism. Data derived primarily from surrogate markers of bone formation indicate that androgens may stimulate osteoblast function. For example, some investigators have reported that serum osteocalcin levels increase when androgens are administered to men (Young et al. 1993) though others have not detected a significant change (Tenover 1992).

Although it appears likely that androgens stimulate osteoblast activity, this finding does not explain the increase in bone resorption that is seen both in animals and men after orchiectomy. Biochemical markers of bone turnover (serum osteocalcin levels and urinary hydroxyproline excretion) increase in men after castration (Stepan et al. 1989). Testosterone administration reduces urinary hydroxyproline excretion in men with borderline low serum testosterone levels (Tenover 1992). Both osteoclast number and the extent of osteoclast-covered bone surfaces increase after orchiectomy and these effects can be prevented by administration of testosterone or dihydrotestosterone. The mechanism by which androgens inhibit the upregulation of osteoclastogenesis that follows orchiectomy may involve effects on local production of cytokines in bone. For example, it has been reported that androgens inhibit the transcription of interleukin-6 (IL-6) and that IL-1 production was increased in a hypogonadal male. Because these cytokines promote osteoclast activation and differentiation, increased local production of IL-1, IL-6, or other osteoclast-stimulating cytokines may be the mechanism whereby androgen deficiency stimulates bone resorption.

Androgens may also affect bone metabolism by effects on calcium regulatory hormones. It has been reported that serum calcitonin levels are lower in hypogonadal men than in normal men; that testosterone administration increases calcitonin levels in hypogonadal men; and that testosterone administration enhances the hypocalcemic effect of calcitonin in orchiectomized rats. One group of investigators reported that testosterone replacement increases serum 1,25-$(OH)_2$ vitamin D levels in hypogonadal men (Francis et al. 1986). However, we did not detect any change in 1,25-$(OH)_2$ vitamin D levels in GnRH-deficient men before or after androgen replacement (Finkelstein et al. 1989) and no changes were observed in 1,25-$(OH)_2$ vitamin D levels of older men with borderline low serum testosterone levels during testosterone therapy (Tenover 1992). Furthermore, no changes in serum 25-OH vitamin D, 1,25-$(OH)_2$ vitamin D, or PTH levels occur during GnRH analog-induced hypogonadism in men (Goldray et al. 1993).

It is still unclear whether the ability of testosterone to stimulate bone formation and inhibit bone resorption is due to testosterone itself or one of its metabolites such as estradiol or dihydrotestosterone. It has been demonstrated that testosterone can be converted to dihydrotestosterone by human bone in vitro. Nonetheless, inhibition of dihydrotestosterone formation by administration of an inhibitor of 5α-reductase has no effect on bone mass in humans (Matzkin et al. 1992) or rats. Several observations suggest that aromatization of testosterone into estrogen is crucial for many of the effects of testosterone on bone. First, estrogen receptors have been demonstrated in human osteoblasts. Secondly, estrogens can maintain bone mass in castrated male-to-female transsexuals. Thirdly, and most importantly, it has recently been reported that a male with complete estrogen resistance due to a genetic defect in the estrogen receptor has severe osteopenia despite normal testosterone levels and complete virilization (Smith et al. 1994). This finding provides the most compelling evidence to date that estrogens are required for a normal peak bone mass in men. However, because cortical bone mineral density is higher in normal men than in women (Bonjour et al. 1991), it appears likely that androgens have an independent effect on peak bone mass. Further studies are needed to assess the relative roles of androgens and estrogens in bone metabolism in men.

7.3.2 Therapy of Androgen-Deficiency Bone Loss

As noted above, several studies have demonstrated increases in bone density in hypogonadal men receiving androgen replacement therapy, particularly in those men who are still skeletally immature (Finkelstein et al. 1989; Greenspan et al. 1989). However, the degree of androgen deficiency needed to produce osteoporosis in men is currently unknown. This issue is of considerable clinical importance because many men are seen with mildly decreased or low normal levels of testosterone and it is not clear whether these men are at risk for developing osteoporosis that might be preventable with testosterone therapy. There are also no data comparing different modes of androgen replacement (e.g., parenteral vs topical) on bone density. Finally, the use of nonandrogenic agents to prevent bone loss in hypogonadal men with contraindications to testosterone therapy (e.g., men with prostate cancer receiving GnRH analog therapy or after surgical castration) has not been examined. Because bone resorption is increased in such men, antiresorptive therapy would seem to be logical. In addition, we have recently demonstrated that daily PTH administration, an agent that stimulates bone formation, prevents GnRH analog-induced bone loss in women with endometriosis, in whom estrogen therapy may be contraindicated (Finkelstein et al. 1994). Thus, it seems likely that PTH might also prevent GnRH analog-induced bone loss in men. The use of anabolic agents like PTH, or antiresorptive agents, such as bisphosphonates or calcitonin, to prevent bone loss in hypogonadal men with contraindications to testosterone therapy needs further evaluation.

References

Bonjour J-P, Theintz G, Buchs B, Slosman D, Rizzoli R (1991) Critical years and stages of puberty for spinal and femoral bone mass accumulation during adolescence. J Clin Endocrinol Metab 73:555–563

Finkelstein JS (1995) Osteoporosis. In: Bennett JC, Plum F (eds) Cecil textbook of medicine. Saunders, Philadelphia (in press)

Finkelstein JS, Klibanski A, Neer RM, Greenspan SL, Rosenthal DI, Crowley WF (1987) Osteoporosis in men with idiopathic hypogonadotropic hypogonadism. Ann Intern Med 106:354–461

Finkelstein JS, Klibanski A, Neer RM, Doppelt SH, Rosenthal DI, Segre GV, Crowley WF (1989) Increases in bone density during treatment of men with idiopathic hypogonadotropic hypogonadism. J Clin Endocrinol Metab 69:776–783

Finkelstein JS, Neer RM, Biller BMK, Crawford JD, Klibanski A (1992) Osteopenia in adult men with histories of delayed puberty. N Engl J Med 326:600–604

Finkelstein JS, Klibanski A, Schaefer EH, Hornstein MD, Schiff I, Neer RM
(1994) Parathyroid hormone for the prevention of bone loss induced by estrogen deficiency. N Engl J Med 331:1618–1623

Francis RM, Peacock M, Aaron JE, Selby PL, Taylor GA, Thompson J, Mar-
shall DH, Horsman A (1986) Osteoporosis in hypogonadal men: role of decreased plasma 1,25-dihydroxyvitamin D, calcium malabsorption, and low
bone formation. Bone 7:261–268

Goldray D, Weisman Y, Jaccard N, Merdler C, Chen J, Matzkin H (1993) Decreased bone density in elderly men treated with the gonadotropin-releasing
hormone agonist decapeptyl (D-Trp6-GnRH). J Clin Endocrinol Metab
76:288–290

Greenspan SL, Neer RM, Ridgway EC, Klibanski A (1986) Osteoporosis in
men with hyperprolactinemic hypogonadism. Ann Intern Med 104:777–782

Greenspan SL, Oppenheim DS, Klibanski A (1989) Importance of gonadal
steroids to bone mass in men with hyperprolactinemic hypogonadism. Ann
Intern Med 110:526–531

Kasperk CH, Wergedal JE, Farley JR, Linkhart TA, Turner RT, Baylink DJ
(1989) Androgens directly stimulate proliferation of bone cells in vitro. Endocrinology 124:1576–1578

Katznelson L, Finkelstein J, Baressi C, Klibanski A (1994) Increase in trabecular bone density and altered body composition in androgen replaced hypogonadal men. Endocrine Society, 76th annual meeting 581 (abstract
1524)

Krabbe S, Christiansen C (1984) Longitudinal study of calcium metabolism in
male puberty. I. Bone mineral content, and serum levels of alkaline phosphatase, phosphate, and calcium. Acta Paediatr Scand 73:745–749

Matzkin H, Chen J, Welsman Y, Goldray D, Pappas F, Jaccard N, Braf Z
(1992) Prolonged treatment with finasteride (a 5α-reductase inhibitor) does
not affect bone density and metabolism. Clin Endocrinol (Oxf) 37:432–436

Seeman E, Melton LJ, O'Fallon WM, Riggs BL (1983) Risk factors for spinal
osteoporosis in men. Am J Med 75:977–983

Smith EP, Boyd J, Frank GR, Takahashi H, Cohen RM, Specker B, Williams
TC, Lubahn DB, Korach KS (1994) Estrogen resistance caused by a mutation in the estrogen-receptor gene in a man. N Engl J Med 331:1056–1061

Stepan JJ, Lachman M, Zverina J, Pacovsky V, Baylink DJ (1989) Castrated
men exhibit bone loss: effect of calcitonin treatment on biochemical indices
of bone remodeling. J Clin Endocrinol Metab 69:523–527

Tenover JS (1992) Effects of testosterone supplementation in the aging male. J
Clin Endocrinol Metab 75:1092–1098

Young NR, Baker HWG, Liu G, Seeman E (1993) Body composition and
muscle strength in healthy men receiving testosterone enanthate for contraception. J Clin Endocrinol Metab 77:1028–1032

8 General Principles of Vitamin D Action and Mechanism-Based Search for Analogs with Specific Actions

H. F. DeLuca, C. Zierold, and H. M. Darwish

8.1   General Introduction 137
8.2   Classical Actions of 1,25-(OH)2D3 138
8.3   The Differential Action of 1,25-(OH)2D3 and the Suppression
        of Cell Growth 141
8.4   Molecular Mechanism of Action of 1,25-(OH)2D3 143
8.5   Search for Analogs with Selective Activity 147
8.6   Selection of Analogs that Are Organ Selective 149
8.7   Evidence for Selective Gene Activation 150
8.8   Summary and Conclusions 153
References 153

8.1 General Introduction

There is no doubt that vitamin D which is normally formed in skin or obtained in the diet must be altered before it can carry out its functions (DeLuca 1974, 1988). First, 25-hydroxylation is carried out in the microsomes and mitochondria of liver to produce the circulating form of vitamin D 25-hydroxyvitamin D_3 (25-OH-D_3). Secondly, 1α-hydroxylation produces the final hormonal or active form 1,25-dihydroxyvitamin D_3 (1,25-$(OH)_2D_3$). In normal humans and animals, this occurs exclusively in the proximal convoluted tubule cells of the kidney, with the notable exception of the placenta (DeLuca 1974, 1988). There is abundant evidence to support the idea that 1,25-dihydroxyvitamin D_3

$(1,25\text{-}(OH)_2D_3)$ is the metabolically active form of vitamin D (DeLuca 1974, 1988). It is believed to carry out both the classical functions of the vitamin and some of the more recently found functions in differentiation and development and in suppression of the parathyroid glands (Darwish and DeLuca 1993; DeLuca 1988, 1992). Furthermore, new and unknown functions of vitamin D will likely be discovered, as, for example, in the female reproductive system (Halloran and DeLuca 1980; Kwiecinski et al. 1989), in the islet cells of the pancreas (Chertow et al. 1983), and in the keratinocytes of skin (Smith et al. 1986). This list will probably become longer and will include abnormal sites such as in malignant tissue that contains significant amounts of vitamin D receptor (VDR) (Eisman 1984).

In addition to the above functions, which are believed to be via a nuclear receptor gene activation mechanism, nongenomic actions of $1,25\text{-}(OH)_2D_3$ have also been postulated (Hruska et al. 1988; Nemere and Norman 1987). So far, these reports have not been convincing. For this reason and because the nongenomic actions often involve large and unphysiologic amounts of $1,25\text{-}(OH)_2D_3$ in vitro, this review will not consider that concept.

8.2 Classical Actions of $1,25\text{-}(OH)_2D_3$

The best known and classical functions of vitamin D are to elevate plasma calcium and plasma phosphorus to levels that are required for formation of bone on the one hand, and to prevent the neuromuscular dysfunction known as hypocalcemic tetany on the other (Darwish and DeLuca 1993; DeLuca 1974, 1988, 1992). Figure 1 illustrates the classical sites of action of $1,25\text{-}(OH)_2D_3$ on the organs which are responsible for elevating plasma calcium and phosphorus concentrations: $1,25\text{-}(OH)_2D_3$ stimulates the enterocyte of the small intestine to transport calcium against an electrochemical potential gradient (Darwish and DeLuca 1993; DeLuca 1992). Via an independent mechanism, it also stimulates the enterocyte to transport inorganic phosphate against an electrochemical potential gradient into the plasma compartment (Darwish and DeLuca 1993; DeLuca 1992). As far as what is known today, these two mechanisms do not involve any other hormone once $1,25\text{-}(OH)_2D_3$ has been formed.

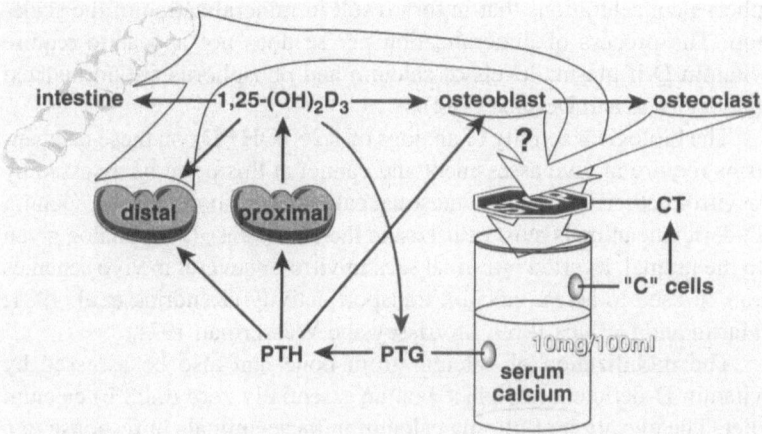

Fig. 1. Regulation of serum calcium by the vitamin D endocrine system. The low serum calcium monitor is the parathyroid gland (*PTG*) that secretes parathyroid hormone (*PTH*). This hormone binds to the entire length of the nephron, and in the proximal tubule stimulates synthesis of 1,25-(OH)$_2$D$_3$, whereas in the distal segment, it, together with parathyroid hormone, stimulates renal reabsorption of calcium. 1,25-(OH)$_2$D$_3$ stimulates intestinal absorption of calcium and phosphorus and stimulates the osteoblast to secure mobilization of bone calcium either directly or through the osteoclast. This process is blocked by calcitonin (*CT*) which is secreted by the C cells of the thyroid in response to high blood calcium. The *heavy arrows* indicate the flow of calcium to raise serum calcium concentration

The 1,25-(OH)$_2$D$_3$ form also plays an important role in the mobilization of calcium from bone (Darwish and DeLuca 1993; DeLuca 1974, 1988, 1992). It is the result of activation of the osteoblast which either directly responds by pumping calcium back into the plasma compartment or by providing a signal to osteoclasts that resorb bone (Suda et al. 1992a). The process of mobilizing calcium from bone in vivo requires the presence of the parathyroid hormone (PTH) (Garabedian et al. 1974). Parathyroidectomy eliminates the action of 1,25-(OH)$_2$D$_3$ on this system, whereas the administration of PTH restores this response (Garabedian et al. 1974). Similarly, in the distal renal tubule, calcium is reabsorbed under the influence of both 1,25-(OH)$_2$D$_3$ and PTH (Yamamoto et al. 1984). Calcium and phosphorus entering the circulation via these mechanisms cause the elevation of plasma calcium and phosphorus concentrations that in turn result in mineralization of the skeleton. The process of mineralization per se does not appear to require vitamin D if plasma levels of calcium and phosphorus are normalized (Underwood and DeLuca 1984).

The biological activity of analogs of 1,25-(OH)$_2$D$_3$ on these mechanisms require in vivo assessment and cannot at this point be assessed by in vitro methods. To assess intestinal calcium transport activity, vitamin D-deficient animals must be used and the 1,25-(OH)$_2$D$_3$ or analog given to the animal. Everted intestinal sacs in vitro or several in vivo schemes can be used to assess calcium transport activity (Kendrick et al. 1981; Martin and DeLuca 1969; Morrissey and Wasserman 1971).

The mobilization of calcium from bone can also be assessed by vitamin D-deficient rats placed on an essentially zero (0.02%) calcium diet. The elevation of plasma calcium in these animals in response to a dose of 1,25-(OH)$_2$D$_3$ or an analog is at the expense of bone. Obviously, the only source of calcium that can be used for this elevation is bone calcium (Blunt et al. 1968; Carlsson 1952). These two organ-specific actions can, therefore, be easily separated at the physiologic level.

In addition to these actions, 1,25-(OH)$_2$D$_3$ can directly stimulate osteoclastic-mediated bone resorption through the osteoblast in organ cultures (Raisz et al. 1972). In these cultures, 1,25-(OH)$_2$D$_3$ or its analogs can act directly without PTH being present (Stern et al. 1983). This system probably measures the role of 1,25-(OH)$_2$D$_3$ in initiating bone remodeling sites and certainly represents one test system that can be used in regard to initiation of bone remodeling sites.

The suppression of the parathyroid glands can be tested both in vivo and in vitro. As will be described below, there is a vitamin D-responsive element in the preproparathyroid gene to which the VDR binds to suppress expression of the preproparathyroid gene (Demay et al. 1992; Silver et al. 1986). In addition, 1,25-(OH)$_2$D$_3$ can actually suppress parathyroid gland size in vitro and in vivo (Silver 1992; Silver et al. 1985). This is an important regulatory mechanism because it tests whether an analog can be used to suppress the secondary hyperparathyroidism which results from renal failure (Silver 1994; Slatopolsky et al. 1984). Organ selectivity may be based on the genes which are expressed in that particular organ in response to the hormone or it may be due to metabolic inactivation in one cell type versus another. This has not yet been determined but remains a distinct possibility for the selection of such analogs.

8.3 The Differential Action of 1,25-(OH)$_2$D$_3$ and the Suppression of Cell Growth

With the availability of high-specific-activity radiolabeled 1,25-(OH)$_2$D$_3$ came the discovery that this hormone localizes in the nuclei of cells not previously appreciated as targets of vitamin D action (Stumpf et al. 1981, 1982). In Table 1 is a list of now proven and putative sites of action of 1,25-(OH)$_2$D$_3$. Most important is the finding by Abe and Suda that 1,25-(OH)$_2$D$_3$, when added to cultures of promyelocytes, induces them to differentiate to monocytes and causes growth suppression (Abe et al. 1981; Tanaka et al. 1982). Similar observations were made with other cancer cell lines (Dokoh et al. 1982; Eisman et al. 1989). Suda et al. (1992b) have continued to study these developmental events and have shown that 1,25-(OH)$_2$D$_3$ causes development of the giant osteoclasts through this promyelocytic lineage. The role of 1,25-(OH)$_2$D$_3$ in the differentiation of several cell types is now well established. It is interesting that osteoclasts can form in abundance in vitamin D-deficient animals, as, for example during lactation (Holtrop et al. 1986; Miller et al. 1982), indicating that 1,25-(OH)$_2$D$_3$ is not exclusively required for this differentiation and more than one pathway of osteoclast formation might exist. The addition of vitamin D compounds topically simulates the in vitro differentiation of keratinocytes, leaving little

Table 1. Cells or tissues that are either known to be targets of 1,25-(OH)$_2$D$_3$ action or those that are expected from the presence of either receptor or nuclear localization of 1,25-(OH)$_2$D$_3$

Proven	Putative
1. Intestinal enterocyte	1. Islet cell – pancreas
2. Osteoblast	2. Endocrine cells – stomach
3. Distal renal cells	3. Pituitary cells
4. Parathyroid cells	4. Ovarian cells
5. Keratinocytes of skin	5. Placenta
6. Promyelocytes, monocytes	6. Epididymis
7. Lymphocytes	7. Brain (hypothalamus)
8. Colon enterocytes	8. Myoblasts (developing)
9. Shell gland	9. Mammary epithelium
10. Chick chorioallantoic membrane	10. Aortic endothelial cells
	11. Skin fibroblasts

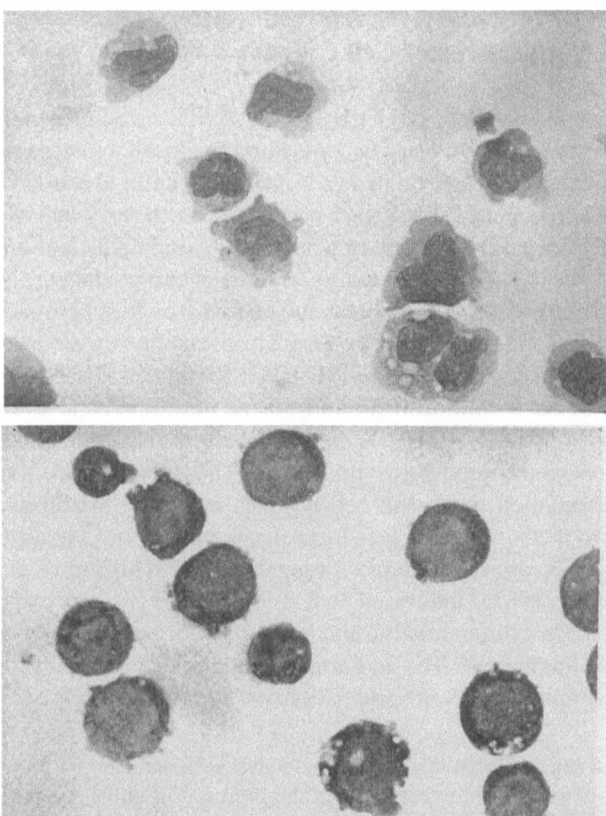

Fig. 2. The differentiation of the myeloid human cell line, HL-60, to the mono-cyte in response to 1,25-(OH)2D3. In the *lower panel* are the undifferentiated HL-60 cells treated with vehicle only (ethanol) and in the *upper panel* are those same cells incubated with 100 nM 1α,25-(OH)2D3 (monocytes). Note the low cytoplasm to nuclear volume ratio in the promyelocytic cells (*lower panel*) and the increased amount of cytoplasm observable in the differentiated cells

doubt about the physiologic significance of this activity (Holick et al. 1987). Figure 2 demonstrates the differentiation of promyelocytic HL-60 cells to monocytes by the addition of 100 nM 1,25-(OH)$_2$D$_3$. The effectiveness, therefore, of the vitamin D compounds in causing suppression of growth and differentiation is another measure of vitamin D function and can be used as an organ- or tissue-specific test for the vitamin D analogs.

8.4 Molecular Mechanism of Action of 1,25-(OH)$_2$D$_3$

The human and rat receptors for 1,25-(OH)$_2$D$_3$ were cloned in 1988 (Baker et al. 1988; Burmester et al. 1988). Figure 3 shows the structure of the VDRs and the domains attributable to these receptors. We recently completed cloning the chicken and Japanese quail receptor and have found it to be a larger protein, primarily by additions to the N-terminal sequence of about 21 amino acids (Elaroussi et al. 1994). In addition, the 3' zinc finger has three different amino acids from the rat

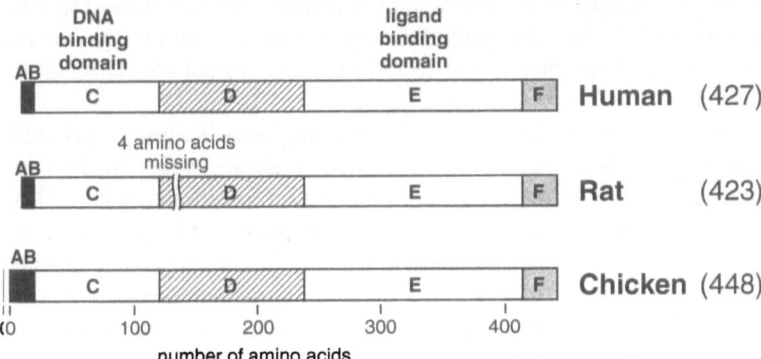

Fig. 3. The known vitamin D receptors. On the *right* in *parentheses* are the number of amino acids deduced from the coding sequence. The major difference in size between the avian and mammalian receptors is an increased number of amino acids in the A/B binding domain. In addition, in the avian species there are three amino acid differences in the most 3' of the two zinc fingers found in the cDNA binding domain. The rat receptor is shorter by 4 amino acids than the human receptor and the missing amino acids are found as indicated in the D domain. *A–D*, binding domains

Table 2. The vitamin D responsive elements found in target genes

Gene	Sequence	Position
CaBP 9K	GGGTGT...AAGCCC	−488 to −474
Rat osteocalcin	GGGTGA...AGGACA	−456 to −442
Human osteocalcin	GGGTGA...GGGGCA	−511 to −486
Mouse osteopontin	GGTTCA...GGTTCA	−757 to −743
Rat 24-OHase distal	GGTTCA...GGTGCG	−262 to −238
Human 24-OHase distal	AGTTCA...GGTGTG	−293 to −273
Rat 24-OHase proximal	GAGTCA...AGGTGA...AGGGCG	−151 to −125
Human 24-OHase proximal	GAGTCA...AGGTGA...AGGGCG	−171 to −143
Suppression		
Human PTH	NNNNNN TGAACCT	−106 to −100

PTH, parathyroid hormone; CaBP, calcium-binding protein.

or human DNA-binding domain which is 100% homologous in those two species. So far, a single human receptor appears to be responsible for all of the actions of 1,25-$(OH)_2D_3$ (Goto et al. 1992). The human receptor has been cloned from HL-60 cells on three different occasions (a cell that undergoes differentiation in response to 1,25-$(OH)_2D_3$) and was found to be identical with the sequence determined by Baker et al. for the human receptor found in intestine and fibroblasts (Goto et al. 1992).

By means of reporter gene constructs and DNA binding or gel shift experiments, the existence of specific response sequences in the promoter region of target genes have been discovered (Ross et al. 1994). These are termed 'vitamin D responsive elements' (VDREs); a list of the most important ones are shown in Table 2. The response elements are not identical and, in fact, differ markedly from gene to gene, including only one gene where the VDR acts in a suppressive mode, i.e., the preproparathyroid gene (Ross et al. 1994).

Of particular interest is the most recent arrival among the VDREs, namely the VDRE system found in the rat and human 1,25-$(OH)_2D_3$–24-hydroxylase genes (Ohyama et al. 1994; Zierold et al. 1994, 1995). These are a powerful responsive element system that imparts a very large response to reporter gene constructs. The fact that the response elements differ widely suggests that it is, indeed, possible that genes differ in their response to analogs, as will be discussed subsequently.

By gel shift analysis, it is possible to demonstrate very clearly that at physiologic concentrations the VDR does not bind to the response elements but rather requires a nuclear accessory protein (Ross et al. 1992; Sone et al. 1991). The RXR class of proteins has been clearly shown to serve as the nuclear accessory factor forming heterodimers between the VDR, the RXR protein, and the response element (Kliewer et al. 1992; Yu et al. 1991). In our laboratory we have isolated in highly purified form the nuclear accessory factor (NAF) from porcine intestine and have found that this protein binds 9-*cis*-retinoic acid (Munder et al. 1995). The isolated NAF serves as an accessory factor for the retinoic acid response elements and retinoic acid receptor γ (RARγ) (Munder et al. 1995). An anti-RXR antibody will combine with both VDR-DRE-NAF complexes, demonstrating that an RXR is a component of both complexes (Munder et al. 1995). There is little doubt that RXR serves as an important co-effector in binding of VDR to the VDREs. The complex formation does not require the presence of the ligand, although in the case of osteocalcin VDRE in the presence of ligand, the complex appears to be more stable to higher salt concentrations (Ross et al. 1993). Thus, current results suggest that ligand is not involved in the formation of the complex between VDR, RXR, and VDRE.

The importance of VDR phosphorylation to the target gene response is not entirely resolved. However, in chick organ cultures, it can be shown that the VDR proteins do not become phosphorylated until the ligand is added (Brown and DeLuca 1990). Then phosphorylation occurs very rapidly and is one of the earliest events found in the response of that tissue to 1,25-$(OH)_2D_3$. This is a serine phosphorylation in the ligand binding domain (Brown and DeLuca 1991). In a reporter gene system, activation of protein kinase A with 8-bromo-cAMP results in a VDR and VDRE-dependent increase in transactivation (Darwish et al. 1993). Furthermore, blocking of phosphorylation by inhibitors of protein kinase A block the reporter gene response to the VDR and hormone. This phosphorylation of VDR is in the ligand-binding domain (Brown and DeLuca 1991), somewhere around amino acid 205 (Hilliard et al. 1994). If this site is mutated, alternate phosphorylation occurs on adjacent sites, obviating conclusions on its essentiality for transactivation (Hilliard et al. 1994). In any case, phosphorylation of the VDR in the ligand-binding domain in response to analog could be a mechanistic site of screening and determination of activity by an analog. A proposed

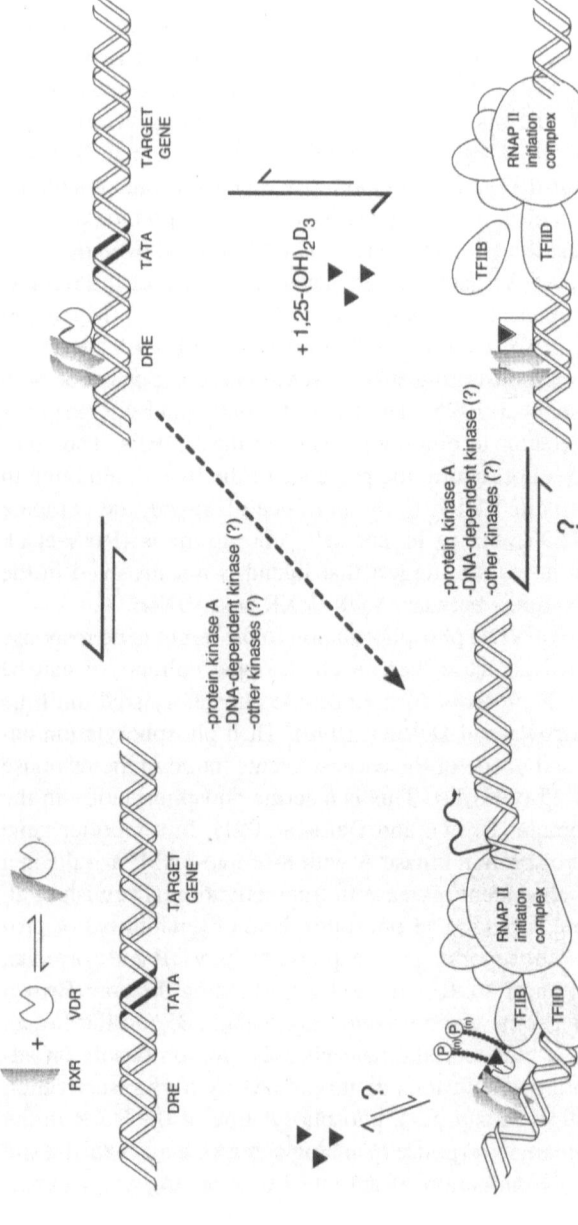

Fig. 4. Our current view of how 1,25-(OH)2D3 and its receptor activates gene transcription. *VDR*, vitamin D receptor; *DRE*, vitamin D response element; *TFIIB*, transcription factor IIB; *TFIID*, transcription factor IID; *RNAP II*, RNA polymerase II; *P*, phosphorus; (*n*), number of phosphorus groups

mechanism whereby 1,25-$(OH)_2D_3$ activates gene transcription encompassing VDR phosphorylation and RXR is shown in Fig. 4.

8.5 Search for Analogs with Selective Activity

Initially, during the course of a routine examination of new metabolites of vitamin D by our group, we found that 24-homo-1,25-$(OH)_2D_3$ and 26-homo-1,25-$(OH)_2D_3$ showed increased activity above 1,25-$(OH)_2D_3$ in the stimulation of HL-60 differentiation to monocytes (Ostrem et al. 1987). On the other hand, 24-homo-1,25-$(OH)_2D_3$ showed a tenfold drop in activity in intestinal calcium transport and in the mobilization of calcium from bone (Perlman et al. 1990). On closer examination, 24-homo-1,25-$(OH)_2D_3$ actually was five to ten times more active than 1,25-$(OH)_2D_3$ in the differentiation assay. This defied the results obtained on receptor binding in which 24-homo-1,25-$(OH)_2D_3$ was found to be 30 times less active than 1,25-$(OH)_2D_3$ in binding to the receptor (Ostrem et al. 1987; Perlman et al. 1990). It became evident that simply studying binding of ligand to the receptor was not by itself of much value in determining in vivo biological activity. As a result of this work, we and other groups proceeded to study side chain modification of the vitamin D molecule in terms of its action on raising blood calcium in vivo and in the in vitro determination of its ability to cause differentiation and growth suppression. This selection was only an end result and did not address the mechanism involved. Clearly, 24-homologation resulted in increased activity in differentiation, a reduction in binding to the receptor, and a marked reduction in calcemic activity (Fig. 5). The major problem is that the discrimination might be the result of reduced binding of the analog to the vitamin D transport protein (Bouillon et al. 1991) and thus rapid in vivo metabolism. Even if this proved to be the mechanism, the rapidly metabolized compounds could be used topically to stimulate keratinocytes to differentiate without consequent systemic hypercalcemic activity (Binderup and Bramm 1988). This undoubtedly accounts for the selective activity of the MC903 compound (Calcipotriol) of Leo Pharmaceuticals (Copenhagen, Denmark; Kragballe et al. 1991), the 22-oxa-1,25-$(OH)_2D_3$ compound of Chugai Pharmaceuticals (Tokyo, Japan) (Abe et al. 1987; Nishii 1994), and the 24-homologated compounds of the Wisconsin group

BASE COMPOUND	COMPOUND	HL-60 ED$_{50}$(M)	Ca^{++} u/mg	HL-60/Ca^{++} Ratio
	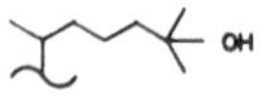	10^{-8}	4×10^4	1
	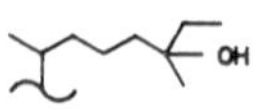	10^{-9}	40×10^4	1
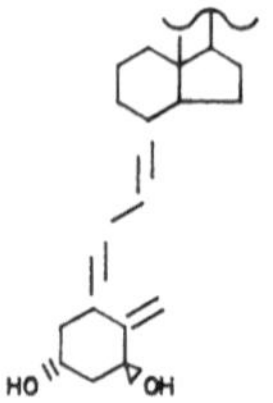	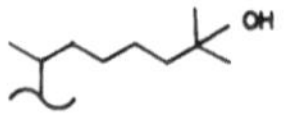	10^{-9}	0.4×10^4	100
	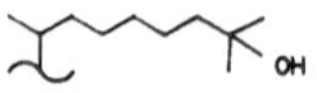	10^{-9}	~400	1000
	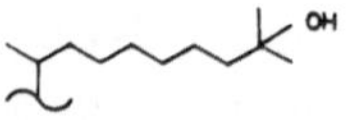	5×10^{-8}	~4	50,000

Fig. 5. Alteration in the side chain structure of 1,25-(OH)₂D₃ and its effects on HL-60 cell differentiation or mobilization of calcium from either intestine or bone. At the *far right* is a ratio of these two activities, taking a ratio of 1 for 1,25-(OH)₂D₃. The differentiation end point is indicated by ED₅₀ or the molar concentration of the indicated compound required for 50% differentiation of the HL-60 cells. Calcium-mobilizing activity is units of activity per milligram of compound. Obviously 24-homologation of the side chain causes a marked reduction in calcium-mobilizing activity and a slight increase in differentiation activity, resulting in a marked change in ratio of the two activities

(Department of Biochemistry, The University of Wisconsin-Madison, USA) (Perlman et al. 1990). Similarly, 26,27-hexafluoro-16-ene-23-yne-1,25-(OH)$_2$D$_3$ and 16-ene-1,25-(OH)$_2$D$_3$ of Hoffmann-La Roche would fall into this category (Anzano et al. 1994; Uskokovic et al. 1991). Thus, the reduction in calcemic activity for whatever reason and the maintenance of differentiative and growth suppressive activity makes possible the use of these compounds for selective activity and treatment of certain diseases. For example, calcipotriol is currently marketed for the treatment of psoriasis (Kragballe et al. 1991) and 22-oxa- and 19-nor-1,25-(OH)$_2$D$_2$ compounds are being developed for the treatment of renal osteodystrophy (Slatopolsky 1994, 1995). The latter two compounds do not raise blood calcium while suppressing the parathyroid glands and PTH secretion. Calcipotriol is effective at the cellular level on topical application but is rapidly cleared once it appears in the blood stream (Jones 1994).

8.6 Selection of Analogs that Are Organ Selective

During the course of studying vitamin D_2 metabolism, 24-epi-1,25-$(OH)_2D_2$ and 1,25-$(OH)_2D_2$ were chemically synthesized (Sicinski et al. 1985). When these compounds were tested in vivo on the mobilization of calcium from bone and intestinal calcium transport, some very surprising results were obtained (DeLuca et al. 1988). Both 24-epi-1,25-$(OH)_2D_2$ and 1,25-$(OH)_2D_2$ were equal to 1,25-$(OH)_2D_3$ in binding to the VDR (Sicinski et al. 1985) and were equal in activity in causing cellular differentiation (Ostrem et al. 1987). However, in vivo, 24-epi proved to be about four to five times less biologically active than 1,25-$(OH)_2D_2$ in intestinal calcium transport. However, 24-epi proved to have virtually no ability to mobilize calcium from bone even at very high doses (DeLuca et al. 1988). This resulted in a series of analogs modified in the 24-position which lack activity in the mobilization of calcium from bone. Thus, organ selectivity can be achieved by analog modification. The mechanism for discrimination can very easily be cell-specific metabolism. As illustrated in the time course of response to bone calcium mobilization, the 24-epi-1,25-$(OH)_2D_2$ does have short-lived activity while it has long-term activity in intestine (DeLuca et al. 1988). This may be the result of more rapid metabolism of 24-epi-1,25-$(OH)_2D_2$ by bone cells or it may be that these analogs can in some way preferentially activate one gene versus another. In any case, the preparation of 1,25-$(OH)_2D_2$ derivatives that lack ability to mobilize calcium from bone while retaining other activities such as bone formation or intestinal calcium transport provide interesting compounds for the treatment of bone loss diseases such as osteoporosis. A series of such compounds are shown in Fig. 6.

Besides the side-chain modifications described above, the 2β-isopropoxy or 2β-benzyloxy-19-nor-1,25-$(OH)2D3$ compounds also show similar organ discrimination (Sicinski et al. 1994).

Structure	Ca Transport	Bone Ca^{++} Mobilization	Bone Formation
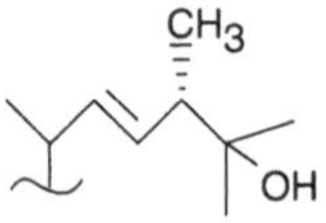	1/1	1/2	1/1
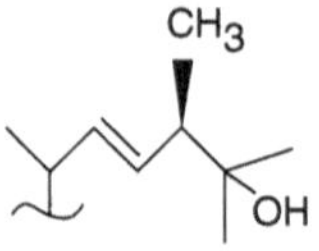	1/4	1/1,000	2/1
	1/2	1/1,000	2/1

Fig. 6. A series of 1,25-(OH)$_2$D$_2$ derivatives and their relative activities in intestinal calcium transport and bone calcium mobilization. The structure depicted is only the side chain which attaches to carbon-17 of the D-ring as shown in Fig. 5. The activity ratio is computed as compared to 1,25-(OH)$_2$D$_3$. Note that 1,25-(OH)$_2$D$_2$ is equal in intestinal calcium transport to 1,25-(OH)$_2$D$_3$ but is one half as active in the mobilization of calcium from bone. Placing the methyl carbon in the epi-position on carbon-24 of 1,25-(OH)$_2$D$_3$ results in a profound change in the ratio of intestinal calcium transport to bone calcium mobilization. The last two compounds show virtually no bone calcium-mobilizing activity in vivo

8.7 Evidence for Selective Gene Activation

Gene selective activity of vitamin D analogs can also be demonstrated. Following the discovery of 24-dihomo-1,25-(OH)$_2$D$_3$, we attempted to learn whether this compound showed equal discrimination on all genes in the same cells in vivo (Krisinger et al. 1991). Doses of this compound were administered to vitamin D-deficient rats at a level 30 times that of 1,25-(OH)$_2$D$_3$ because its binding to the VDR is 1/30th that of 1,25-(OH)$_2$D$_3$. A comparison of the dose response of the intestinal calbindin D-9k gene and intestinal calcium transport is provided in Fig. 7. In the same cell, the 24-homologated analogs did not increase intestinal calcium transport, whereas it produced the expected response in the calbin-

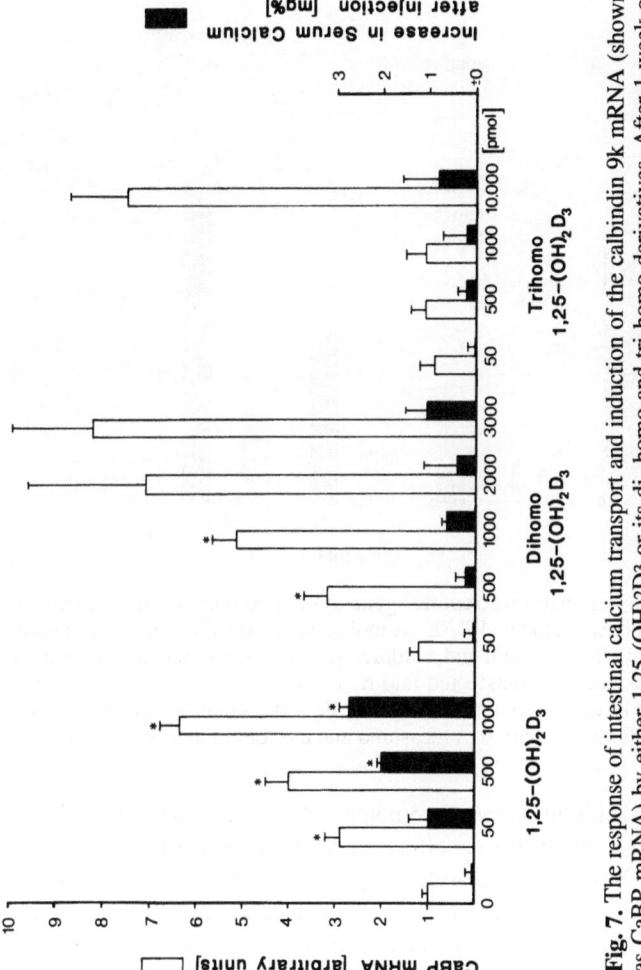

Fig. 7. The response of intestinal calcium transport and induction of the calbindin 9k mRNA (shown as CaBP mRNA) by either 1,25-(OH)2D3 or its di- homo and tri-homo derivatives. After 1 week of chronic doses, the intestinal calcium transport and calbindin mRNA were measured. The homologated compounds possess virtually no activity in intestinal calcium transport while being similar to 1,25-(OH)2D3 in inducing the calbindin 9k mRNA. The dihomo compound binds 1/30 as well to the receptor and the trihomo compound binds 1/130th as well as does 1,25-(OH)2D3

din D-9k according to comparison with 1,25-(OH)$_2$D$_3$. In the same cell, where presumably metabolism is identical, one gene is stimulated and other genes that are responsible for calcium transport are not, illustrating a mechanism-related discrimination.

When reporter constructs were made of the human osteocalcin DRE placed in front of a chloramphenicol acetyltransferase reporter gene

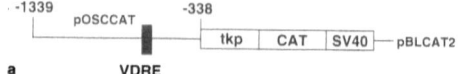

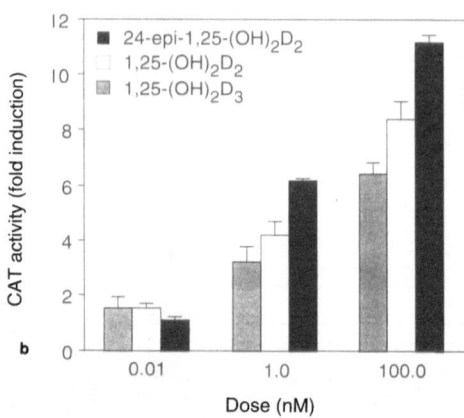

Fig. 8a,b. The response of a reporter gene system containing the osteocalcin vitamin D response element (*VDRE*) as indicated in (**a**) spliced in the promoter of a thymidylic kinase system and a chloramphenical acetyltransferase reporter gene. This plasmid was transfected into rat osteosarcoma 17/2.8 cells and the indicated dose of the compounds was given. Four days later the chloramphenical acetyltransferase activity was measured and the fold induction computed

having a thymidylic kinase promoter, it is clear that 24-epi-1,25-(OH)$_2$D$_3$ had a much greater effect on reporter gene activity than did either 1,25-(OH)$_2$D$_2$ or 1,25-(OH)$_2$D$_3$ (Fig. 8). In northern analysis of the osteocalcin transcript levels in osteosarcoma cells, 24-epi-1,25-(OH)$_2$D$_2$ stimulates the osteocalcin gene much better than 1,25-(OH)$_2$D$_3$ (Arbour et al. 1995). As described above, however, this very same compound had a much lower activity on the mobilization of calcium from bone which undoubtedly involves genes expressed in osteoblasts. There is very likely to be a mechanism-based differential response of target genes to analogs. These results indicate that 24-modified compounds might show preferential activity for anabolic genes in bone cells and, thus, may be useful in restoring bone that has been lost because of diseases such as osteoporosis.

8.8 Summary and Conclusions

Unfortunately, an insufficient amount of information is available on the molecular mechanism whereby $1,25\text{-}(OH)_2D_3$ stimulates gene expression. Certainly the binding of ligand to the receptor does not appear to play a role in the binding of VDR to the response elements. Yet the response elements appear different for each responsive gene discovered so far. This may mean that the response elements are involved beyond merely binding of the receptor. Improved action of certain response elements by analogs can be demonstrated, illustrating that the overall mechanism is in some way related to the response elements. Additionally, there may be organ-based discrimination among analogs and metabolic-based discrimination, all of which can be useful in the design of analogs of vitamin D to specifically treat some diseases with a minimum of side effects. However, it is very clear that insufficient information is available concerning how $1,25\text{-}(OH)_2D_3$ causes gene expression and, furthermore, there is an insufficient amount of information on the genes which are required to produce the phenotype brought about by the actions of vitamin D. Once some of these facts are known and once the three-dimensional structure of the VDR is known, analogs for specific functions can be designed and screened on a more rational basis.

Acknowledgments. This work was supported in part by a program project grant no. DK14881 from the National Institutes of Health, a fund from the National Foundation for Cancer Research, and a fund from the Wisconsin Alumni Research Foundation.

References

Abe E, Miyaura C, Sakagami H, Takeda M, Konno K, Yamazaki T, Yoshiki S, Suda T (1981) Differentiation of mouse myeloid leukemia cells induced by $1\alpha,25$-dihydroxyvitamin D_3. Proc Natl Acad Sci USA 78:4990–4994

Abe J, Morikawa M, Miyamoto K, Kaiho S, Fukushima M, Miyaura C, Abe E, Suda T, Nishii Y (1987) Synthetic analogues of vitamin D_3 with an oxygen atom in the side chain skeleton: a trial of the development of vitamin D compounds which exhibit potent differentiation-inducing activity without inducing hypercalcemia. FEBS Lett 226:58–61

Anzano MA, Smith JM, Uskokovic MR, Peer CW, Mullen LT, Letterio JJ, Welsh MC, Shrader MW, Logsdon DL, Drive CL et al (1994) 1α,25-Dihydroxy-16-ene-23-yne-26,27-hexafluorocholecalciferol (Ro24-5531), a new deltanoid (vitamin D analogue) for prevention of breast cancer in the rat. Cancer Res 54(7):1653–1656

Arbour NC, Darwish HM, DeLuca HF (1995) Transcriptional control of the osteocalcin gene by 1,25-dihydroxyvitamin D_2 and its 24-epimer in rat osteosarcoma cells. Biochim Biophys Acta (in press)

Baker AR, McDonnell DP, Hughes M, Crisp TM, Mangelsdorf DJ, Haussler MR, Pike JW, Shine J, O'Malley BW (1988) Cloning and expression of full-length cDNA encoding human vitamin D receptor. Proc Natl Acad Sci USA 85:3294–3298

Binderup L, Bramm E (1988) Effects of a novel vitamin D analogue MC 903 on cell proliferation and differentiation in vitro and on calcium metabolism in vivo. Biochem Pharmacol 37:889–891

Blunt JW, Tanaka Y, DeLuca HF (1968) The biological activity of 25-hydroxycholecalciferol, a metabolite of vitamin D_3. Proc Natl Acad Sci USA 61:1503–1506

Bouillon R, Allewaert K, Xiang DZ, Tan BK, Baelen HV (1991) Vitamin D analogs with low affinity for the vitamin D binding protein: enhanced in vitro and decreased in vivo activity. J Bone Miner Res 6:1051–1056

Brown TA, DeLuca HF (1990) Phosphorylation of the 1,25-dihydroxyvitamin D_3 receptor: a primary event in 1,25-dihydroxyvitamin D_3 action. J Biol Chem 265:10025–10029

Brown TA, DeLuca HF (1991) Sites of phosphorylation and photoaffinity labeling of the 1,25-dihydroxyvitamin D_3 receptor. Arch Biochem Biophys 286:466–472

Burmester JK, Wiese RJ, Maeda N, DeLuca HF (1988) Structure and regulation of the rat 1,25-dihydroxyvitamin D_3 receptor. Proc Natl Acad Sci USA 85:9499–9502

Carlsson A (1952) Tracer experiments on the effect of vitamin D on the skeletal metabolism of calcium and phosphorus. Acta Physiol Scand 26:212–220

Chertow BS, Sivitz WI, Baranetsky NG, Clark SA, Waite, A, DeLuca HF (1983) Cellular mechanisms of insulin release. The effects of vitamin D deficiency and repletion on rat insulin secretion. Endocrinology 113:1511–1581

Darwish HM, Burmester J, Moss VE, DeLuca HF (1993) Phosphorylation is involved in transcriptional activation by the 1,25-dihydroxyvitamin D_3 receptor. Biochim Biophys Acta 1167:29–36

Darwish H, DeLuca HF (1993) Vitamin D regulated gene expression. In: Stein GS, Stein JL, Lian JB (eds) Critical reviews in eukaryotic gene expression, vol 3(2). CRC Press, Boca Raton, pp 89–116

DeLuca HF (1974) Vitamin D: the vitamin and the hormone. Fed Proc 33:2211–2219

DeLuca HF (1988) The vitamin D story: a collaborative effort of basic science and clinical medicine. FASEB J 2:224–236

DeLuca HF (1992) New concepts of vitamin D functions. In: Sauberlich HE, Machlin KJ (eds) Beyond deficiency. New views on the function and health effects of vitamins, vol 669. New York Academy of Sciences, New York, pp 59–69

DeLuca HF, Sicinski R, Tanaka Y, Stern PH, Smith CM (1988) The biological activity of 1,25-dihydroxyvitamin D_2 and 24-epi-1,25-dihydroxyvitamin D_2. Am J Physiol 17:E402–E406

Demay MB, Kiernan MS, DeLuca HF, Kronenberg HM (1992) Sequences in the human parathyroid hormone gene that bind the 1,25-dihydroxyvitamin D_3 receptor and mediate transcriptional repression in response to 1,25-dihydroxyvitamin D_3. Proc Natl Acad Sci USA 89:8097–8101

Dokoh S, Donaldson CA, Haussler MR (1984) Influence of 1,25-dihydroxyvitamin D_3 on cultured osteogenic sarcoma cells: correlation with the 1,25-dihydroxyvitamin D_3 receptor. Cancer Res 44:2103–2109

Eisman JA (1984) 1,25-Dihydroxyvitamin D3 receptor and role of 1,25-(OH)2D3 in human cancer cells. In: Kumar R (ed) Vitamin D, chap 14. Nijhoff, Boston, pp 365–382

Eisman JA, Koga M, Sutherland RL, Barkla DH, Tutton PJM (1989) 1,25-Dihydroxyvitamin D_3 and the regulation of human cancer cell replication. Proc Soc Exp Biol Med 191:221–226

Elaroussi MA, Prahl JM, DeLuca HF (1994) The avian vitamin D receptors: primary structures and their origins. Proc Natl Acad Sci USA 91:11596–11600

Garabedian M, Tanaka Y, Holick MF, DeLuca HF (1974) Response of intestinal calcium transport and bone calcium mobilization to 1,25-dihydroxyvitamin D_3 in thyroparathyroidectomized rats. Endocrinology 94:1022–1027

Goto H, Chen K-S, Prahl JM, DeLuca HF (1992) A single receptor identical with that from intestine/T47D cells mediates the action of 1,25-dihydroxyvitamin D-3 in HL-60 cells. Biochim Biophys Acta 1132:103–108

Halloran BP, DeLuca HF (1980) Effect of vitamin D deficiency on fertility and reproductive capacity in the female rat. J Nutr 119:1573–1580

Hilliard GM, IV, Cook RC, Weigel NL, Pike JW (1994) 1,25-Dihydroxyvitamin D_3 modulates phosphorylation of serine 205 in the human vitamin D receptor: site-directed mutagenesis of this residue promotes alternative phosphorylation. Biochemistry 33:4300–4311

Holick MF, Smith E, Pincus S (1987) Skin as the site of vitamin D synthesis and target tissue for 1,25-dihydroxyvitamin D_3: use of calcitriol (1,25-dihydroxyvitamin D_3) for treatment of psoriasis. Arch Dermatol 123:1677–1679

Holtrop ME, Cox KA, Carnes DL, Holick MF (1986) Effects of serum calcium and phosphorus on skeletal mineralization in vitamin D-deficient rats. Am J Physiol 251:E234–E240

Hruska KA, Bar-Shavit Z, Malone JD, Teitelbaum S (1988) Ca^{2+} priming during vitamin D-induced monocytic differentiation of a human leukemia cell line. J Biol Chem 263:16039–16045

Jones G (1994) Vitamin D target cells degrade 1α,25-dihydroxyvitamin D_3 (1a,25-$(OH)_2D_3$) and its analogues into inactive side-chain shortened metabolites. In: Sida T (ed) Vitamin D and its analogues. The 2nd international forum on calcified tissue and bone metabolism, 4–5 November 1993. Chugai Pharmaceuticals, Tokyo, pp 30–33

Kendrick NC, Kabakoff B, DeLuca HF (1981) Oxygen-dependent 1,25-dihydroxycholecalciferol-induced calcium ion transport in rat intestine. Biochem J 194:178–186

Kliewer SA, Umesono K, Mangelsdorf DJ, Evans RM (1992) Retinoid X receptor interacts with nuclear receptors in retinoic acid, thyroid hormone and vitamin D_3 signaling. Nature 355:446–449

Kragballe K, Gjertsen BT, DeHoop D, Karlsmark T, Van de Kerkhof PCM, Larko O, Nieboer C, Roed-Petersen J, Strand A, Tikjob G (1991) Double-blind, right/left comparison of calcipotriol and βmethasone valerate in treatment of psoriasis vulgaris. Lancet 337:193–196

Krisinger J, Strom M, Darwish HD, Perlman K, Smith C, DeLuca HF (1991) Induction of calbindin-D 9k mRNA but not calcium transport in rat intestine by 1,25-Dihydroxyvitamin D_3 24-homologs. J Biol Chem 266:1910–1913

Kwiecinski GG, Petrie GI, DeLuca HF (1989) 1,25-Dihydroxyvitamin D_3 restores fertility of vitamin D-deficient female rats. Am J Physiol 256:E483–E487

Martin DL, DeLuca HF (1969) Calcium transport and the role of vitamin D. Arch Biochem Biophys 134:139–148

Miller SC, Halloran BP, DeLuca HF, Jee WSS (1982) Role of vitamin D in maternal skeletal changes during pregnancy and lactation: a histomorphometric study. Calcif Tissue Int 34:245–252

Morrissey RL, Wasserman RH (1971) Calcium absorption and calcium-binding protein in chicks on differing calcium and phosphorus intakes. Am J Physiol 220:1509–1515

Munder M, Herzberg IM, Zierold C, Moss VE, Hanson K, Clagett-Dame M, DeLuca HF (1995) Identification of the porcine intestinal accessory factor that enables DNA sequence recognition by vitamin D receptor. Proc Natl Acad Sci USA 92:2795–2799

Nemere I, Norman AW (1987) Rapid action of 1,25-dihydroxyvitamin D_3 on calcium transport in perfused chick duodenum: effect of inhibitors. J Bone Miner Res 2:99–103

Nishii Y (1994) Future prospects for vitamin D analogues: a overview. In: Suda T (ed) Vitamin D and its analogues. The 2nd international forum on calcified tissue and bone metabolism, 4–5 Nov 1993. Chugai Pharmaceuticals, Tokyo, pp 42–47

Ohyama Y, Ozono K, Uchida M, Shinki T, Kato S, Suda T, Yamamoto O, Noshiro M, Kato Y (1994) Identification of a vitamin D-responsive element in the 5'-flanking region of the rat 25-hydroxyvitamin D_3 24-hydroxylase gene. J Biol Chem 269:10545–10550

Ostrem VK, Lau WF, Lee SH, Perlman K, Prahl J, Schnoes HK, DeLuca HF, Ikekawa N (1987) Induction of monocytic differentiation of HL-60 cells by 1,25-dihydroxyvitamin D analogs. J Biol Chem 262:14164–14171

Perlman K, Kutner A, Prahl J, Smith C, Inaba M, Schnoes HK, DeLuca HF (1990) 24-Homologated 1,25-dihydroxyvitamin D_3 compounds: separation of calcium and cell differentiation activities. Biochemistry 29:190–196

Raisz, LG, Trummel CL, Holick MF, DeLuca HF (1972) 1,25-dihydroxycholecalciferol: a potent stimulator of bone resorption in tissue culture. Science 175:768–769

Ross TK, Moss VE, Prahl JM, DeLuca HF (1992) A nuclear protein essential for binding of rat 1,25-dihydroxyvitamin D_3 receptor to its response elements. Proc Natl Acad Sci USA 89:256–260

Ross TK, Darwish HM, Moss VE, DeLuca HF (1993) Vitamin D-influenced gene expression via a ligand-independent, receptor-DNA complex intermediate. Proc Natl Acad Sci USA 90:9257–9260

Ross TK, Darwish HM, DeLuca HF (1994) Molecular biology of vitamin D action. In: Litwack G (ed) Vitamins and hormones, vol 49. Academic, San Diego, pp 281–326

Sicinski RR, Tanaka Y, Schnoes HK, DeLuca HF (1985) Synthesis of $1\alpha,25$-dihydroxyvitamin D_2, its 24-epimer and related isomers, and their binding affinity for the 1,25-dihydroxyvitamin D_3 receptor. Bioorg Chem 13:158–169

Sicinski RR, Perlman KL, DeLuca HF (1994) Synthesis and biological activity of 2-hydroxy and 2-alkoxy analogs of $1\alpha,25$-dihydroxy-19-nor-vitamin D_3. J Med Chem 37:3730–3738

Silver J (1994) Regulation of parathyroid hormone production by $1\alpha,25$-$(OH)_2D_3$ and its analogues. In: Suda T (ed) Vitamin D and its analogues. The 2nd international forum on calcified tissue and bone metabolism, 4–5 November 1993. Chugai Pharmaceutical, Tokyo, pp 60–63

Silver J, Russell J, Sherwood LM (1985) Regulation by vitamin D metabolites of messenger ribonucleic acid for preproparathyroid hormone in isolated bovine parathyroid cells. Proc Natl Acad Sci USA 82:4270–4273

Silver J, Naveh-Many T, Mayer H, Schmeizer HJ, Popvtzer MM (1986) Regulation by vitamin D metabolites of parathyroid hormone gene transcription in vivo in the rat. J Clin Invest 78:1296–1301

Slatopolsky E (1994) The role of $1\alpha,25$-dihydroxyvitamin D_3 ($1\alpha,25(OH)_2D_3$) and 22-oxacalcitriol (OCT) in suppressing secondary hyperparathyroidism (2° HPT) in chronic renal failure. In: Suda T (ed) Vitamin D and its analogues. The 2nd international forum on calcified tissue and bone metabolism, 4–5 November 1993. Chugai Pharmaceuticals, Tokyo, pp 64–67

Slatopolsky E, Weerts C, Thieland J, Horst R, Harter H, Martin K (1984) Marked suppression of secondary hyperparathyroidism by I.V. administration of 1,25-dihydroxycholecalciferol in uremic patients. J Clin Invest 74:2136–2138

Slatopolsky E, Finch J, Brown A, Wilkins R, DeLuca HF (1994) A new analog of calcitriol, 19-nor-1,25-$(OH)_2D_2$, suppresses PTH secretion in uremic rats in the absence of hypercalcemia. American Society of Nephrology, October 1994, Orlando (abstract)

Smith EL, Walworth NC, Holick MF (1986) Effect of $1\alpha,25$-dihydroxyvitamin D_3 on the morphologic and biochemical differentiation of cultured human epidermal keratinocytes growth in serum-free conditions. J Invest Dermatol 86:709–714

Sone T, Ozono K, Pike JW (1991) A 55-kilodalton accessory factor facilitates vitamin D receptor DNA binding. Mol Endocrinol 5:1578–1586

Stern PH, Halloran BP, DeLuca HF, Hefley TJ (1983) Responsiveness of vitamin D-deficient fetal rat limb bones to parathyroid hormone in culture. Am J Physiol. 244:E421–E424

Stumpf WE, Sar M, DeLuca HF (1981) Sites of action of 1,25 $(OH)_2$ vitamin D_3 identified by thaw-mount autoradiography. In: Cohn DV, Talmage RV, Matthews JL (eds) Hormonal control of calcium metabolism. Excerpta Medica, Amsterdam, pp 222–229

Stumpf WE, Sar M, Clark SA, DeLuca HF (1982) Brain target sites for 1,25-dihydroxyvitamin D_3. Science 215:1403–1405

Suda T, Takahashi N, Abe E (1992a) Role of vitamin D in bone resorption. J Cell Biol 49:53–58

Suda T, Takahashi N, Martin TJ (1992b) Modulation of osteoclast differentiation. Endocr Res 13:66–80

Tanaka H, Abe E, Miyaura C, Kuribayashi T, Konno K, Nishii Y, Suda T (1982) $1\alpha,25$-Dihydroxycholecalciferol and a human myeloid leukaemic cell line (HL-60). The presence of a cytosol receptor and induction of differentiation. Biochem J 204:713–719

Underwood JL, DeLuca HF (1984) Vitamin D is not directly necessary for bone growth and mineralization. Am J Physiol 246:E493–E498

Uskokovic MR, Baggiolini E, Shiuey S-J, Iacobelli J, Hennessy B, Kiegiel J, Daniewski AR, Pizzolato G, Courtney LF, Horst RL (1991) The 16-ene analogs of 1,25-dihydroxycholecalciferol. Synthesis and biological activity. In: Norman AW, Bouillon R, Thomasset M (eds) Vitamin D. Gene regulation, structure-function analysis and clinical application, vol 1. De Gruyter, New York, pp 139–145

Yamamoto M, Kawanobe, Y, Takahashi H, Shimazawa E, Kimura S, Ogata E (1984) Vitamin D deficiency and renal calcium transport in the rat. J Clin Invest 74:507–513

Yu, VC, Delsert C, Andersen B, Holloway JM, Devary OV, Naar AM, Kim SY, Boutin J-M, Glass CK, Rosenfeld MG (1991) RXRβ: a coregulator that enhances binding of retinoic acid, thyroid hormone, and vitamin D receptors to their cognate response elements. Cell 67:1251–1266

Zierold C, Darwish HM, DeLuca HF (1994) Identification of a vitamin D-response element in the rat calcidiol (25-hydroxyvitamin D₃) 24-hydroxylase gene. Proc Natl Acad Sci USA 91:900–902

Zierold C, Darwish HM, DeLuca HF (1995) Two vitamin D response elements function in the rat 1,25-dihydroxyvitamin D 24-hydroxylase promoter. J Biol Chem 270:1675–1678

9 Organ-Specific Actions of Vitamin D Analogs: Relevance of Rapid Effects

M. C. Farach-Carson and S. E. Guggino

9.1   Introduction 162
9.2   Materials and Methods 164
9.2.1   Cell Model and Culture 164
9.2.2   Vitamin D Analogs 164
9.2.3   Calcium Influx Assays 165
9.2.4   Electrophysiology 165
9.2.5   Northern Blotting 166
9.2.6   Transfection Assays 166
9.3   Results 166
9.3.1   Target Cells Recognize Unique Structural Features
        of Vitamin D Metabolites and Analogs 166
9.3.2   Selective Actions of Vitamin D Analogs on Calcium Influx
        Assessed Using Ion Tracer Assays 167
9.3.3   Whole Cell and Single Channel Patch Clamp Recordings
        Demonstrate a Shift in the Threshold of Activation
        of L-Type Calcium Channels by 1,25(OH)$_2$D$_3$
        and 1-Deoxy Analogs and Increase Channel Open Time 169
9.3.4   Northern Blot Analysis Reveals Modulation of mRNA Levels
        Encoding Matrix Proteins by Analogs Which Interact
        with the Nuclear Receptor, but not 1-Deoxy Analogs 172
9.3.5   Transfection Assays Demonstrate Transcriptional Control
        of OPN Biosynthesis by 1,25(OH)$_2$D$_3$ and Analog BT 174
9.4   Discussion 175
References 178

9.1 Introduction

$1,25(OH)_2D_3$ is a calcitropic hormone whose ability to modulate the biological activity and differentiation state of a variety of target cells has long been appreciated (DeLuca 1986; Reichel et al. 1989; Suda et al. 1990). Until recently, it was widely accepted that all of these actions were mediated through specific binding to the nuclear receptor for vitamin D (VDR), a member of the steroid hormone receptor family (Haussler 1986). Nuclear receptor-mediated events generally require many hours, if not days, to produce maximum and long-lasting effects. Examples of physiological processes regulated in this manner by $1,25(OH)_2D_3$ likely include bone matrix modeling and synthesis, cellular growth and differentiation, phenotypic modulation requiring gross changes in patterns of protein biosynthesis, and long-term endocrine effects. Another aspect of the action of $1,25(OH)_2D_3$ which has received considerable attention of late is its ability to stimulate rapid actions in target cells. These rapid actions include stimulation of lipid turnover (Lieberherr et al. 1989; DeBoland and Boland 1993), Ca^{2+} influx via plasma membrane voltage-sensitive Ca^{2+} channels (Caffrey and Farach-Carson 1989; DeBoland and Norman 1990), release of Ca^{2+} from intracellular stores (Civitelli et al. 1990; Lieberherr 1987), and changes in phosphorylation patterns of intracellular and secreted proteins (Safran et al. 1994; Qin et al. 1994). Such rapid effects likely function as cellular signals for secretory "cross-talk" between cells in complex organs, cellular activation, potentiation of biomechanical transduction, and rapid regulation of homeostatic mechanisms involved in mineral metabolism. We believe that the long-term and rapid actions of steroid hormones such as $1,25(OH)_2D_3$ are intimately tied and together account for the pleiotropic effects of $1,25(OH)_2D_3$ on a variety of target cells and organs. Based upon electrophysiological data in our laboratories, we also believe that $1,25(OH)_2D_3$ serves a unique "priming" function at the level of the plasma membrane. This function, observed at low nanomolar concentrations of $1,25(OH)_2D_3$, shifts the threshold of activation of inward calcium currents to more negative and physiological potentials and at the same time leads to prolonged open time of individual channels. One likely consequence of the shift in the activation threshold is an increased responsiveness to other calcitropic hormones, such as parathyroid hormone, which are also known to activate plasma membrane calcium influx (Yamaguchi et al. 1987). Other metabolites of vitamin D, such as $24,25(OH)_2D_3$, also demonstrate plasma membrane effects, but these tend to be inhibitory rather than stimulatory (Caffrey and Farach-Carson 1989; Khoury et al. 1995). The data presented in this article, some of which are new and some of which have been collected from published observations, support our hypothesis that $1,25(OH)_2D_3$ is a unique seco-steroid with a critical role in cellular and systemic calcium homeostasis.

In an effort to dissect the long-term and rapid effects of $1,25(OH)_2D_3$, we have devoted considerable effort to measure the ability of structural analogs of $1,25(OH)_2D_3$ to stimulate various aspects of target cell activation, including both genomic and plasma membrane-mediated events. Although $1,25(OH)_2D_3$ presumably functions as the natural ligand for initiation of both long-term and rapid responses in target cells, we hoped to identify subsets of response pathways that would be activated by discrete structural analogs. This approach would not only allow us to distinguish the response pathways and cellular transducers likely to be involved in activation events in various target cell types, but also allow the development of bioactive compounds related to vitamin D for specific therapeutic use. We have been successful in the dissection of cellular response pathways using structural analogs of $1,25(OH)_2D_3$ (Farach-Carson et al. 1991, 1993), which allowed us to identify bioactive analogs which are either "calcemic" (activate plasma membrane Ca^{2+} signaling events and may or may not be genomically active) or noncalcemic (do not stimulate Ca^{2+} influx). In cases in which they have been studied in whole animals, noncalcemic analogs in vitro also are less likely to produce hypercalcemia in vivo, which may prove to be an important observation for the development of pharmaceuticals (Zhou et al. 1990; Abe et al. 1991; Colston et al. 1992).

9.2 Materials and Methods

9.2.1 Cell Model and Culture

The cells used in these experiments were ROS 17/2.8 osteosarcoma
cells, which display a nonmineralizing osteoblastic phenotype (Majeska
et al. 1980). Cells were cultured as described (Caffrey and Farach-Carson 1989; Yukihiro et al. 1994) in steroid-depleted or serum-free medium. All experiments were performed using growth phase cells, which
consistently demonstrate maximum responsiveness to treatment with
$1,25(OH)_2D_3$. Electrophysiological experiments were performed using
cells replated at low density onto coverslips as previously described
(Caffrey and Farach-Carson 1989; Yukihiro et al. 1994).

9.2.2 Vitamin D Analogs

To remain consistent with our previous reports, we will refer to analogs
in this article using the nomenclature adopted at the University of
California, Riverside (Bouillon et al. 1995). Table 1 lists the structural
description of all analogs referred to in this article. All analogs were

Table 1. Description of metabolites and analogs: UC-Riverside code name,
structural name, other names

UC-Riverside code	Structural name	Other names
C	$1,25(OH)_2D_3$	Calcitriol, vitamin D hormone
BO	$25(OH)_2D_3$	
AS	$24,25(OH)_2D_3$	
BT	$1,24 (OH)_2 -22$-ene-24-cyclopropyl-D_3	Calcipotriol, MC 903
EU	22-oxa-$1,25(OH)_2D_3$	22-oxacalcitriol, OCT
Y	25-(OH)-23-yne-D_3	
HO	25-(OH)-16,23-diene-D_3	
AT	25-(OH)-16-ene-23-yne-D_3	
V	$1,25(OH)_2$-16-ene-23-yne-D_3	
HL	1-$\beta,25(OH)_2D_3$	

stored under nitrogen in the dark at $-20°$. Prior to use, absorption spectra were routinely obtained as described (Farach-Carson et al. 1991), which provided the basis for determination of all concentrations reported.

9.2.3 Calcium Influx Assays

Measurements of transmembrane $^{45}Ca^{2+}$ influx were made using 1-min assays, exactly as described in several of our publications (Caffrey and Farach-Carson 1989; Farach-Carson et al. 1991, 1993). The cellular response to depolarization with elevated levels of extracellular K+ was consistently used to normalize influx data (Caffrey and Farach-Carson 1989; Farach-Carson et al. 1991). This response is maximal in cells from 40%–60% confluency, indicating that ROS 17/2.8 cells in the growth phase exhibit maximal numbers of functional voltage-sensitive Ca^{2+} channels. All influx assays were performed in serum-free medium so that selective interaction with the serum vitamin D-binding protein could be eliminated as a variable. This is especially important for analog comparison, where significant differences in binding to the serum protein have been measured (Bouillon et al. 1991).

9.2.4 Electrophysiology

The solutions and procedures used in patch clamp experiments designed to measure the activity of cell surface inward Ca^{2+} currents in our respective laboratories have been published (Caffrey and Farach-Carson 1989; Yukihiro et al. 1994). Both whole cell and single channel recording configurations have been used. Summed, averaged single channel records show activation and decay kinetics which are virtually identical to the macroscopic currents, confirming that the L-type voltage-sensitive calcium channel is responsible for generating the high threshold macroscopic current seen in whole cell recordings (Caffrey and Farach-Carson 1989). This channel is not the T-type channel identified in primary cultures of neonatal rat calvaria (Chesnoy-Marchais and Fritsch 1988).

9.2.5 Northern Blotting

Isolation of total RNA from growth phase cell monolayers and northern hybridizations were performed as described previously (Farach-Carson et al. 1993; Khoury et al. 1994). All data were recorded from autoradiographs of ^{32}P-hybridized blots exposed to film for 1–2 days. Band intensities corresponding to mRNA encoding the bone matrix proteins osteopontin (OPN) and osteocalcin (OCN) and 28S ribosomal RNA were quantitated densitometrically from scanned autoradiograms (Khoury et al. 1994).

9.2.6 Transfection Assays

The rat OPN gene and 5'-flanking sequences were isolated and sequenced (Ridall et al. 1995). An EcoRI fragment (nucleotides -2318 to -914) was subcloned in the pGL2-pr vector (Promega, Madison, WI, USA) immediately upstream of the SV-40 promoter and a luciferase reporter. This construct contains vitamin D response elements (VDRE) and acts as an enhancer in functional assays (Ridall et al. 1995). As we reported, ROS 17/2.8 cells were successfully transfected with this construct, which serves as a convenient assay for transcriptional activity of analogs or metabolites of vitamin D being tested for transcriptional activity mediated by the nuclear receptor for $1,25(OH)_2D_3$ (Khoury et al. 1994).

9.3 Results

9.3.1 Target Cells Recognize Unique Structural Features of Vitamin D Metabolites and Analogs

In addition to the three naturally occurring metabolites of vitamin D [1, $25(OH)_2D_3$, C; 24, $25(OH)_2D_3$, AS; and the metabolic precursor of both, $25(OH)_2D_3$, BO], our laboratories have together tested a number of structural analogs of $1,25(OH)_2D_3$ from which certain general conclusions concerning their actions on target osteoblasts can be made. The analogs listed in Table 1 were selected from among many others that we have tested which demonstrate unique and measurable actions. We have identified two analogs which bind one third to half as well as C to the nuclear receptor (Farach-Carson et al. 1991, 1993), namely, BT (1,24 $(OH)_2$ –22-ene-24-cyclopropyl-D_3) and EU (22-oxa-1,25$(OH)_2D_3$) and which apparently are selectively able to activate nuclear receptor-mediated pathways, leading to increased gene transcription of matrix proteins. We have identified three analogs lacking the 1-α-hydroxyl group, namely, Y (25-(OH)-3-yne-D_3); HO (25-(OH)-16,23-diene-D_3); and AT (25-(OH)-16-ene-23-yne-D_3) which do not interact with nuclear receptors (Farach-Carson et al. 1991; Bouillon et al. 1995), but which readily activate plasma membrane voltage-sensitive calcium channels (VSCCs) and stimulate influx of extracellular Ca^{2+}(Yukihiro et al. 1994). We have identified others such as V (1,25$(OH)_2$–16-ene-23-yne-D_3) which have moderate effects on all response pathways tested in osteoblasts (Farach-Carson et al. 1991; Khoury et al. 1994). We have also examined the 1-β-isomer of 1,25$(OH)_2D_3$ (HL), which acts as a selective antagonist of plasma membrane-mediated events but does not alter nuclear receptor-mediated responses (Yukihiro et al. 1994; Norman et al. 1993; Baran and Sorensen 1994). For brevity, in all of the following sections we will refer to the various analogs of vitamin D by their two-letter UC-Riverside code names, which are described along with common names, if applicable, in Table 1.

9.3.2 Selective Actions of Vitamin D Analogs on Calcium Influx Assessed Using Ion Tracer Assays

ROS 17/2.8 cells display a resting uptake of $^{45}Ca^{2+}$ that is stimulated approximately threefold in the presence of high (120–132 mM) external K^+ (inset to Figs. 1a,b). Such response in the presence of depolarizing extracellular solutions is characteristic of cells expressing surface VSCCs, and the level of stimulation is directly related to the concentration of functional VSCCs on the cell surface. As shown in the inset to Figs. 1a,b, the native hormone compound C consistently elevates transmembrane $^{45}Ca^{2+}$ influx to a value approximately 75% that seen during complete depolarization. Dose–response curves for $^{45}Ca^{2+}$ influx in the presence of analogs AT and BT are shown in Fig. 1. Analog AT, which is a 1-deoxy derivative of C having a fixed side chain, is active, eliciting

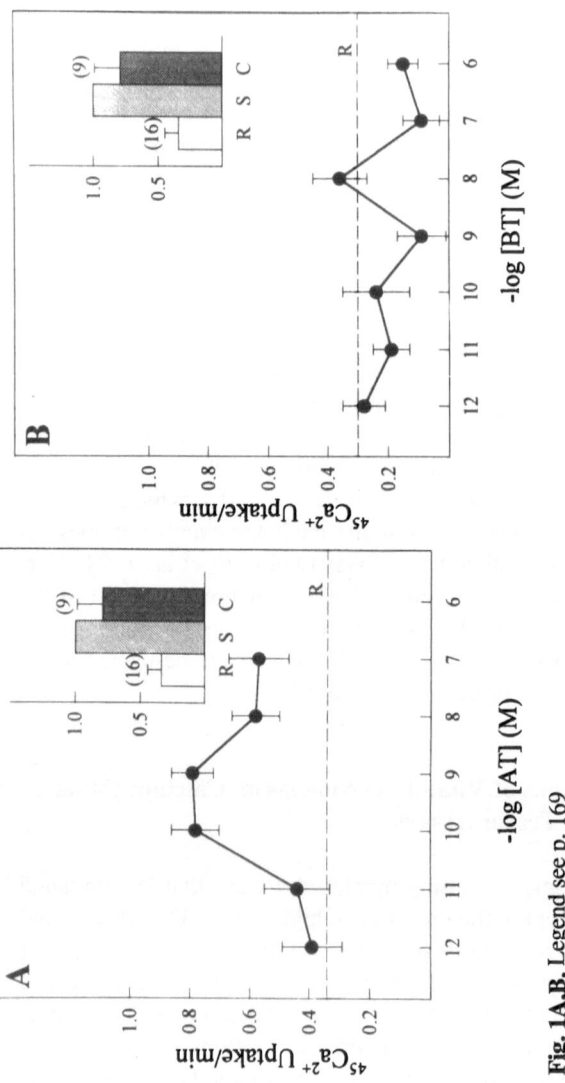

Fig. 1A,B. Legend see p. 169

a maximum response at 0.1 to 1.0 nM (Fig. 1A). In contrast, compound BT demonstrates no stimulatory effect on transmembrane influx of $^{45}Ca^{2+}$ (Fig. 1B). All other analogs tested fell within the range defined by analogs AT and BT, which are the most and least effective analogs, respectively, that we have tested in calcium influx assays. Compound EU is indistinguishable from compound BT; compound V demonstrates moderate stimulation. The exact values for these and other analogs have been published (Farach-Carson et al. 1991, 1993).

9.3.3 Whole Cell and Single Channel Patch Clamp Recordings Demonstrate a Shift in the Threshold of Activation of L-Type Calcium Channels by 1,25(OH)$_2$D$_3$ and 1-Deoxy Analogs and Increase in Channel Open Time

As shown in Fig. 2A and Table 2, we found that 1,25(OH)$_2$D$_3$ and 1-deoxy analogs Y, AT, and HO consistently produced immediate changes in inward Ca^{2+} currents detected using the whole cell recording configuration. These effects were dose dependent (Table 2), and only these analogs, like 1,25(OH)$_2$D$_3$, were effective at low (0.5–1 nM) concentrations. As reported previously (Caffrey and Farach-Carson 1989), the threshold for activation of current was negatively shifted approximately 20 mV, towards the resting potential. These shifts are similar to, but not additive with, the effects of the dihydropyridine VSCC agonist BAY K 8644 (Fig. 2a).

◀ **Fig. 1A,B.** Calcium influx dose responses for analogs AT and BT. Concentration curves for stimulation of transmembrane $^{45}Ca^{+2}$ influx by analogs AT and BT are shown. Various concentrations of each analog from 10^{-12} to 10^{-6} were added from stocks in absolute ethanol to normal K$^+$ resting solution. Experiments were performed as described in Sect. 9.3 "Materials and Methods" to assay rapid (1 min) influx of $^{45}Ca^{+2}$ into growth phase ROS 17/2.8 cells. The *dotted line* represents the level of influx in cells receiving vehicle in resting buffer alone. For comparison, the *inset* to each figure part shows the amount of influx occurring under identical conditions for resting (normal K$^+$) vs. stimulated (high K$^+$). Also shown is the relative level of influx stimulated by inclusion of 1 nM 1,25(OH)$_2$D$_3$. The value in *parentheses* over each *bar* in the inset represents the number of experiments used in the calculation of the mean and standard deviation. *R*, resting; *S*, stimulating; *C*, calcitriol 1,25(OH)$_2$D$_3$ treated

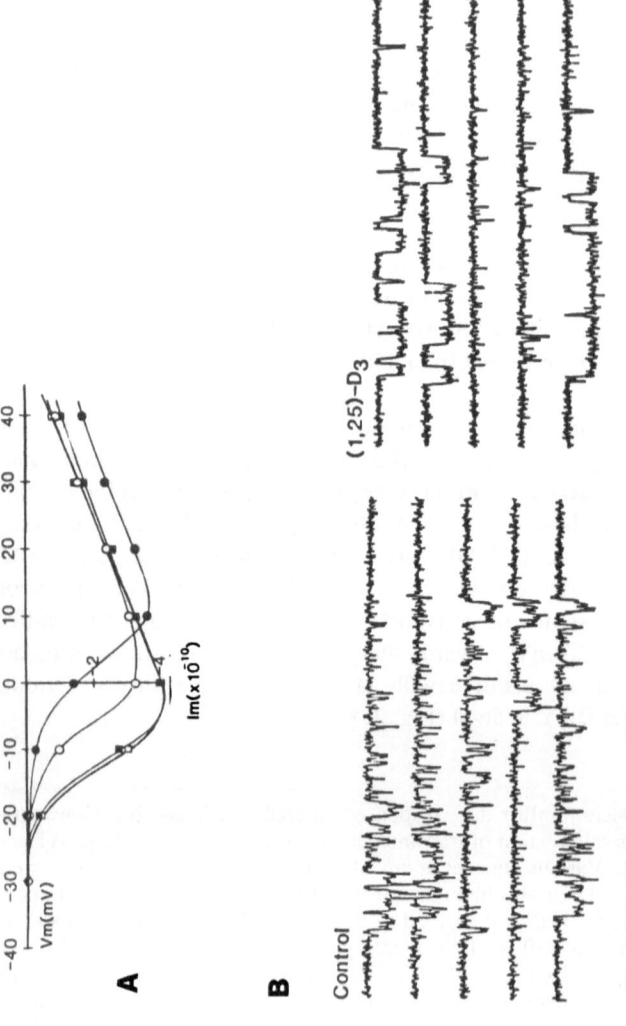

Fig. 2A,B. Legend see p. 171

Table 2. Effects of metabolites and analogs on inward currents through voltage-sensitive calcium channels

Analog/metabolite	Average Peak Shift (mV)		
	0.5 nM	5 nM	50 nM
C	11.9 ± 2	17.3 ± 1.5	19.4 ± 1.4
BO	1.6 ± 1.6	4.7 ± 3.7	14.0 ± 3.6
AS	0.7 ± 0.6	1.4 ± 1.9	2.8 ± 2.2
BT	1.0 ± 0.4	3.7 ± 1.4	10.9 ± 1.8
EU	ND	ND	ND
Y	9.7 ± 2.3	14.3 ± 3.9	$17.0 \pm .7$
AT	9.7 ± 0.3	15.8 ± 1.4	17.9 ± 0.8
HO	14.6 ± 3.6	18.8 ± 0.7	19.8 ± 0.7
V	4.1 ± 1.5	8.5 ± 1.4	12.1 ± 1.6
HL	1.5 ± 1.1	5.2 ± 1.7	8.5 ± 2.3
HL + C	1.5 ± 1.1	5.2 ± 1.7	8.5 ± 2.3

Note: Some of these data are reproduced with permission from Yukihiro et al. (1994)

◀ **Fig. 2A,B.** Whole cell and single channel recordings for $1,25(OH)_2D_3$ showing shift in threshold of activation and prolongation of open time. **A** Current–voltage relations from a ROS 17/2.8 cell under whole cell patch clamp at an extended range of test potentials. *Filled circles,* control relation from a resting osteoblast; *open circles,* following application of 5 nM $1,25(OH)_2D_3$; *open squares,* 1 nM $1,25(OH)_2D_3$. Note the shift of the threshold of activation toward the resting potential following application of the hormone. The *filled squares* illustrate that addition of the calcium channel agonist BAY K8644 subsequent to $1,25(OH)_2D_3$ produces no additional shift of activation or increase of current amplitude. **B** $1,25(OH)_2D_3$ prolongs the open time of unitary Ca^{2+} channels. This panel shows consecutive traces at a test potential of 0 mV, before (*left*) and after (*right*) addition of 3 nM $1,25(OH)_2D_3$. Note the prolongation of channel openings and the increased incidence of traces with low opening probability after application of the hormone. The holding potential in this experiment was –50 V

Further experiments to investigate the mechanism of current stimulation by $1,25(OH)_2D_3$ using single channel recording techniques were performed (Fig. 2b). As shown, addition of 3 nM C to the bath solution of a cell attached patch-induced prolongation of open time during test depolarizations to 0 mV. These consecutive traces also illustrate an increased incidence of traces with lower open probability than control. These changes seen at the single channel level can account completely for the macroscopic current changes, as determined from comparison of summed, averaged single channel currents to actual whole cell recordings (Caffrey and Farach-Carson 1989). It is interesting to note that the 1-β analog of compound C (analog HL) does not alone produce a left shift in channel current, but when added with compound C was able to attenuate the effects of the native hormone (Table 2). This is similar to the effects noted during calcium influx assays reported previously (Norman et al. 1993).

9.3.4 Northern Blot Analysis Reveals Modulation of mRNA Levels Encoding Matrix Proteins by Analogs Which Interact with the Nuclear Receptor, but not 1-Deoxy Analogs

The analogs described here bind with widely differing abilities to the nuclear receptor in solution. These values, based upon competitive binding assays, have been previously reported (reviewed in Bouillon et al. 1995). Of the analogs listed in Table 1, only compounds BT, EU, and V bind with reasonable affinity to the solubilized nuclear receptor from ROS 17/2.8 cells, with relative values to C of 115%, 48%, and 30%, respectively (Farach-Carson et al. 1991, 1993). As readily seen from northern blotting data shown in Table 3, only those analogs which bind effectively to the nuclear receptor (BT,EU,V) were capable of forming transcriptionally active complexes involving the VDR. The relative increases in mRNA levels were slightly different for the two osteoblastic marker proteins, with an increase of 2.4- to 3.4 -fold found for OPN and 3- to 6-fold found for OCN. 1-Deoxy analogs (AT, Y) are not able to induce this increase in VDR-mediated mRNA accumulation, nor is the β-compound HL able to block it (Khoury et al. 1994; Norman et al. 1993). Interestingly, the calcium channel blocker nitrendipine did not block the increase in mRNA levels induced by $1,25(OH)_2D_3$ either

Table 3. Northern blot analysis demonstrates modulation of mRNA levels encoding bone matrix proteins by $1,25(OH)_2D_3$ and analogs BT, V and EU, but not AT or Y

Treatment	Induced/uninduced mRNA level[a]			48-h (veh)[b]	Fold induction[c]
	3-h	24-h	48-h		
OPN					
C	2.8	8.3	7.1	2.1	3.4
BT	1.0	4.3	4.3	1.6	2.6
EU	ND	4.1	3.9	1.6	2.4
AT	1.6	2.3	1.1	1.8	0.6
Y	ND	1.8	2.0	2.0	1.0
V	2.1	3.8	4.0	1.3	3.1
OCN					
C	3.4	24.6	32.0	5.4	5.9
BT	0.7	16.5	17.8	5.4	3.3
EU	ND	17.3	25.6	5.1	5.0
AT	0.6	2.5	2.0	2.8	0.7
Y	ND	ND	ND	ND	ND
V	0.7	4.6	5.2	1.0	5.2

[a] Autoradiographs of northern blots were scanned using a densitometer and the band intensities quantitated. The level of mRNA present at 0 time ("uninduced") was assigned a value of 1.0. All values reported were normalized to the 0 time lane as a ratio of "induced/uninduced" mRNA levels.

[b] Control cultures were grown in parallel in serum-free medium with vehicle (veh) only.

[c] To correct for the background level of mRNA induction seen during culture in serum-free medium (Khoury et al. 1994), the ratio of mRNA at 48 h in treated versus untreated cultures is reported.

(Khoury et al. 1994). It should be noted that compound V, which would be produced if the genomically inactive AT were hydroxylated at the 1-position (Khoury et al. 1994), is capable of binding to the VDR and upregulating levels of both OPN and OCN.

9.3.5 Transfection Assays Demonstrate Transcriptional Control of OPN Biosynthesis by 1,25(OH)$_2$D$_3$ and Analog BT

In vitro transfection assays allow the direct measurement of transcriptional activity of an analog, i.e., a way to measure its ability to functionally interact with specific nucleotide sequences upstream of reporter genes. Using this approach, we used a VDRE-containing promoter–reporter construct derived from the genomic rat OPN sequence to test the ability of analogs AT and BT to stimulate transcription. Compound C was used as the positive control. As shown in Fig. 3, analog BT, but not

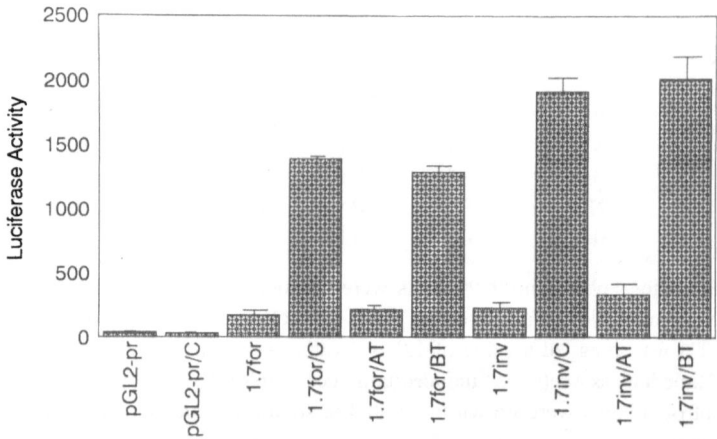

Fig. 3. Transfection assays demonstrating transcriptional control of osteopontin (OPN) biosynthesis by 1,25(OH)$_2$D$_3$ and analogs. Analog BT, but not AT, induces vitamin D responsive elements (VDRE)-mediated transcription of OPN. A construct derived from the sequence of genomic OPN promoter region that contains an active VDRE was used to test the ability of analogs AT and BT to stimulate transcription of a luciferase reporter gene. As the positive control 1,25(OH)$_2$D$_3$ (compound C) was used. ROS 17/2.8 cells at 50% confluency were transfected in triplicate as described in Sect. 9.3 "Materials and Methods." Cells were harvested and luciferase activity measured. Results were normalized using an internal β-galactosidase control. *pGL2-pr*, promoter only; *1.7*, the 1.7-kb restriction fragment containing the VDRE region coupled to the promoter; *for*, forward orientation; *inv*, inverted orientation; *C, AT*, and *BT* structures are described in Table 1. Note that analog BT, but not AT, is able to stimulate VDRE region-mediated transcription in transfected osteoblastic cells

AT, upregulates the reporter expression in an orientation-independent manner, consistent with the enhancer function of the VDRE. These results correlate with their in vitro binding patterns to the osteoblast VDR (BT, 115%; AT, 0.08% compared to C).

9.4 Discussion

The $1,25(OH)_2D_3$ signaling system in target cells such as osteoblasts clearly involves an intricately woven network of cellular events, some of which are initiated at the plasma membrane and some of which are initiated as a consequence of activation of nuclear receptors. Until recently, it has been difficult to catalog these various response pathways because of the phenotypic coupling which occurs in responsive cells. For example, ROS 17/2.8 cells at confluence simultaneously lose both VDR-mediated and membrane-initiated responsiveness to $1,25(OH)_2D_3$ (Farach-Carson et al. 1991; Majeska et al. 1980). Cell lines such as the ROS 24/1, while reportedly containing membrane-initiated pathways without measurable numbers of VDRs (Baran et al. 1991), nonetheless vary phenotypically. For example, we have never detected voltage-dependent inward calcium currents in these cells using the patch clamp (Caffrey and Farach-Carson, unpublished).

A more successful approach has evolved from the design and use of structural analogs of $1,25(OH)_2D_3$ and related metabolites. A very comprehensive review of these analogs was recently published, along with their structural features (Bouillon et al. 1995). Use of these analogs in in vitro assays allows the dissection of the various intracellular pathways affected by $1,25(OH)_2D_3$, and provides experimental tools for the examination of the integrated circuitry in target cells. A practical product of basic investigations using the structural analogs and various target cells as models is the characterization of the "bioprofiles" of individual analogs. This in vitro information then provides a sound basis for selection of analogs for animal studies directed toward therapeutic use in human disease.

At the present time, most potential therapeutic strategies involve the development of analogs which produce beneficial effects mediated through the VDR without producing hypercalcemia, the most obvious and consistent side effect of vitamin D therapy (Berl et al. 1981; Salusky et al. 1987). Analogs which fall into this category include BT, EU, and perhaps V. All three of these analogs, as shown here, interact well with the osteoblast VDR, upregulate mRNA levels encoding the osteoblastic markers OPN and OCN, and stimulate transcription of a reporter construct containing the VDRE of rat OPN. Neither EU nor BT can stimulate influx of Ca^{2+} through VSCCs, but V can stimulate influx about 20%–50% as well as $1,25(OH)_2D_3$. All analogs in this category are expected to bind well to nuclear receptors (Farach-Carson et al. 1991; Bouillon et al. 1995), upregulate target genes containing VDRE upstream sequences (Pike 1991), downregulate production of parathyroid hormone (Russell et al. 1986), modulate 1- and 24-hydroxylase activity (Kumar 1984), and stimulate differentiation of certain target cells, including keratinocytes and osteoclasts (Su et al. 1994; Suda 1989). Limited reports indicate that they do this with diminished tendency to produce hypercalcemia and related side effects (Zhou et al. 1990; Colston et al. 1992; Abe et al. 1991). In fact, BT is now approved for topical use for the treatment of psoriasis, under the trade name Dovonex. It is interesting to speculate that the reduction in the incidence and severity of hypercalcemia produced by these analogs is related to their failure to trigger Ca^{2+} influx into target cells or to shift the activation theshold of VSCCs toward the resting potential. Since we believe that this shift toward the resting potential produced by physiological concentrations of $1,25(OH)_2D_3$ (but not analogs such as BT) can increase responsiveness to other calcitropic hormones, it is reasonable to predict that systemic bone and mineral homeostasis would be shifted toward net resorption by these compounds rather than by the parent compound, $1,25(OH)_2D_3$. Further experimentation will be needed to clearly prove this prediction.

The therapeutic usefulness of compounds such as AT, HO, and Y, which stimulate VSCCs and produce the left shift in VSCC activation but do not activate nuclear receptor-mediated pathways, is less clear at this time. Perhaps their utility will evolve as they are subjected to further study, including whole animal challenge. Nonetheless, these analogs already provide important experimental tools for the study of membrane-initiated events induced by the seco-steroids. They have proven useful in pharmacological analyses of membrane receptors for vitamin D related compounds (Nemere 1994) and may well be the means for the isolation and definative identification of plasma membrane seco-steroid receptors.

The final type of analog is illustrated by the 1-β compound HL, which acts as a selective antagonist of plasma membrane-initiated but not VDR-mediated cellular events (Yukihiro et al. 1994; Norman et al. 1993; Baran and Sorensen 1994). If membrane-initiated responses to 1,25(OH)$_2$D$_3$ are proven to correlate with hypercalcemic tendency, analogs of this type may gain therapeutic value in conditions in which there is minimum endocrine compromise, but in which the metabolic conditions have been shifted toward resorption. An example of this situation would be the astronaut during extended spaceflight or the bed-rested fracture patient, who may be producing normal amounts of renal 1,25(OH)$_2$D$_3$, but for whom the absence of mechanical stimulation has led to bone resorption and elevated free calcium. Again, further work will be needed to investigate these predictions.

It is becoming increasingly clear that vitamin D and its "cousins" in the steroid hormone family play vital roles in the regulation of many physiological processes. These include both rapid (membrane-initiated) and long-term (nuclear receptor-mediated) pathways in various target cells. Research in this rapidly growing field is on the threshold of revealing some fundamental concepts of cellular integration and responsiveness to endocrine and paracrine signals. Vitamin D analogs provide not only valuable tools for basic research, but also candidate compounds for future drug development.

Acknowledgements. The authors are grateful to all of the members of our laboratories whose assistance and ideas have supported this investigation. We are indebted to Dr. Anthony Norman, whose long-term collaboration has been essential for the development and success of this project. Dr. Farach-Carson wishes to especially thank Ms. Laura Adams and Ms. Donna Duron for their support with grant applications and manuscripts, Mr. Gerald Bellot for technical assistance, Dr. Norman Karin for many useful discussions, and Mr. Jeff Safran for his assistance with graphics. This work was supported by grants from the National Institutes of Health (DE-10318), to M.C.F.-C. and DK-43423, (to S.E.G.).

References

Abe J Nakano T, Nishii Y, Matsumoto T, Ogatea E, Ikeda K (1991) A novel vitamin D_3 analog, 22-oxa-1,25-dihydroxyvitamin D_3, inhibits the growth of human breast cancer in vitro and in vivo without causing hypercalcemia. Endocrinology 129:832–837

Baran DT, Sorensen AM (1994) Rapid actions of 1,25-dihydroxyvitamin D_3 physiologic role. Proc Soc Exp Biol Med 207:175–179

Baran DT, Sorensen AM, Shalhoub V, Owen T, Oberdorf A, Stein G, Lian J (1991) 1,25-Dihydroxyvitamin D_3 rapidly increases cytosolic calcium in clonal rat osteosarcoma cells lacking the vitamin D receptor. J Bone Miner Res 6:1269–1275

Berl T, Berns AS, Huffler WE, Hammill AC, Alfrey AC, Arnaud CD, Shrier RW (1981) 1,25-Dihydroxycholecalciferol effects in chronic dialysis: a double blind controlled study. Ann Intern Med 88:774–780

Bouillon R, Allewaert K, Xiang DZ, Tan BK, van Baelen H (1991) Vitamin D analogs with low affinity for the vitamin D binding protein: enhanced in vitro and decreased in vivo activity. J Bone Miner Res 6:1051–1057

Bouillon R, Okamura WH, Norman (1995) Structure-function relationships in the vitamin D endocrine system. Endocr Rev 16:200–257

Caffrey JM, Farach-Carson MC (1989) Vitamin D_3 metabolites modulate dihydropyridine-sensitive calcium currents in clonal rat osteosarcoma cells. J Biol Chem 264:20265–20274

Chesnoy-Marchais D, Fritsch J (1988) Voltage-gated sodium and calcium currents in rat osteoblasts. J Physiol (Lond) 398:291–311

Civitelli R, Kim YS, Gunsten SL, Fujimori A, Huskey M, Avioli LV Hruska KA (1990) Nongenomic activation of the calcium message system by vitamin D metabolites in osteoblast-like cells. Endocrinology 127:2253–2262

Colston KW, Chander SK, Mackay AG, Coombes RC (1992) Effects of synthetic vitamin D analogues on breast cancer cell proliferation in vivo and in vitro. Biochem Pharmacol 44:693–702

DeBoland AR, Boland RL (1993) 1,25-Dihydroxyvitamin D_3 induced arachidonate mobilization in embryonic chick myoblasts. Biochem Biophys Acta (Mol Cell Res) 1179:98–104

De Boland AR, Norman AW (1990) Influx of extracellular calcium mediates 1,25-dihydroxyvitamin D_3-dependent transcaltachia. Endocrinology 127:2475–2480

DeLuca HF (1986) The metabolism and functions of vitamin D. Adv Exp Med 196:361–375

Farach-Carson MC, Sergeev I, Norman AW (1991) Nongenomic actions of 1,25-dihydroxyvitamin D_3 in rat osteosarcoma cells: structure-function studies using ligand analogs. Endocrinology 129:1876–1884

Farach-Carson MC, Abe J, Nishii Y, Khoury R, Wright GC, Norman AW (1993) 22-Oxacalcitriol: dissection of 1,25(OH)$_2$D$_3$ receptor mediated and Ca^{2+} entry-stimulating pathways. Am J Physiol 265 (Renal Fluid Electrolyte Physiol 34):F705–F711

Haussler MR (1986) Vitamin D receptors: nature and function. Annu Rev Nutr 6:527–562

Kumar R (1984) Metabolism of 1,25-dihydroxyvitamin D3. Physiol Rev 64:478–504

Khoury R, Ridall AL, Norman AW, Farach-Carson MC (1994) Target gene activation by 1,25-dihydroxyvitamin D$_3$ in osteosarcoma cells is independent of calcium influx. Endocrinology 135:2446–2453

Khoury R, Weber J, Farach-Carson MC (1995) Vitamin D metabolites modulate osteoblast activity by Ca^{2+} influx-independent genomic and Ca^{2+} influx-dependent nongenomic pathways. J Nutr 125:1699S–1703S

Lieberherr M (1987) Effects of vitamin D metabolites on cytosolic free calcium in confluent mouse osteoblasts. J Biol Chem 262:13168–13173

Lieberherr M, Grosse B, Duchambon P, Drueke T (1989) A functional cell surface type receptor is required for the early action of 1,25-dihydroxyvitamin D$_3$ on the phosphoinositide metabolism in rat enterocytes. J Biol Chem 264:20403–20406

Majeska RJ, Rodan SB, Rodan GA (1980) Parathyroid hormone-responsive clonal cell lines from rat osteosarcoma. Endocrinology 107:1494–1503

Nemere I (1994) Membrane receptors for 1,25(OH)$_2$D$_3$ and 24,25(OH)$_2$D$_3$ act through different signal transduction pathways as deduced by effects on intestinal phosphate transport. Mol Biol Cell 5:18a

Norman AW, Bouillon R, Farach-Carson MC, Bishop JE, Zhou L-X Nemere I, Zhao DJ, Muralidharan KR, Okamua WH (1993) Demonstration that 1β,25-dihydroxyvitamin D$_3$ is an antagonist of the nongenomic but not genomic biological responses and biological profile of the three A-ring diastereomers of 1α,25-dihydroxyvitamin D$_3$. J Biol Chem 268:20022–20030

Pike JW (1991) Vitamin D3 receptors: structure and function in transcription. Annu Rev Nutr 11:189–216

Qin X, Siaw WKO, Walters MR (1994) 1,25-Dihydroxyvitamin D$_3$ effects in rat kidney: regulation of protein phosphorylation. Biochem Biophys Res Commun 204:807–812

Reichel H, Koeffler HP, Norman AW (1989) The role of the vitamin D endocrine system in health and disease. N Engl J Med 320:980–991

Ridall AL, Dickinson DP, Daane EL, DeLuca HF, Butler WT (1995) Characterization of the rat osteopontin gene: evidence for two vitamin D response elements. In: Denhardt DT (ed) Osteopontin: role in cell signalling and adhesion. Ann NY Acad Sci 760:59–66

Russell J, Lettieri D, Sherwood LM (1986) Suppression by 1,25-(OH)$_2$D$_3$ of transcription of the parathyroid hormone gene. Endocrinology 119:2864–2866

Safran JB, Wright GC, Khoury R, Butler WT, Farach-Carson MC (1994) Alteration of charge state of the bone matrix protein osteopontin (OPN) induced by 1,25(OH)$_2$D$_3$. 9th Workshop on Vitamin D. In: Norman AW Bouillon R, Thomasset M (eds) Vitamin D, a pluripotent steroid hormone: structural studies, molecular endocrinology, and clinical applications. De Gruyter, Berlin

Salusky IB, Fine RS, Kangarloo H, Gold L, Paunier WG, Goodman JE, Brill JE, Gilli G, Slatopolsky E, Coburn JW (1987) "High-dose" calcitriol for control of renal osteodystrophy in children on CAPD. Kidney Int 32:89–95

Su MJ, Bikle DD, Mancianti ML, Pillai S (1994) 1,25-Dihydroxyvitamin D$_3$ potentiates the keratinocyte response to calcium. J Biol Chem 269:14723–14729

Suda T (1989) The role of 1,25-dihydroxyvitamin D$_3$ in the myeloid cell differentiation. Proc Soc Exp Biol Med 191:214–220

Suda T, Shinki T, Takahashi N (1990) The role of vitamin D in bone and intestinal cell differentiation. Annu Rev Nutr 10:195–211

Yamaguchi DT, Hahn TJ, Iida-Klein A, Kleeman CR, Muallem S (1987) Parathyroid hormone-activated calcium channels in an osteoblast-like clonal osteosarcoma cell line. J Biol Chem 262:7711–7718

Yukihiro S, Posner GH, Guggino SE (1994) Vitamin D$_3$ analogs stimulate calcium currents in rat osteosarcoma cells. J Biol Chem 269:23889–23893

Zhou JY, Norman AW, Chen DL, Sun GW, Uskokovic M, Koeffler HP (1990) 1,25-Dihydroxy-16-ene-23-yne-vitamin D$_3$ prolongs survival time of leukemic mice. Proc Natl Acad Sci USA 87:3929–3932

Subject Index

adrenal cortex   57
androgens   122–123, 132–134
antiestrogen   30–35, 42
antiprogestins   107–108, 110, 116

breast cancer   86–87, 93, 95–97, 101

calcium channels   167, 169, 171
cell proliferation   85–92, 95–97, 99–100, 102
chromaffin cells   54, 57–58
contraception   113
cyclic AMP   29, 33, 37, 40–44
cyclin   96–97, 99–102
cyclin-dependent kinase   96, 100

delayed puberty   123, 130, 132

endometrial development   108, 110–111, 113–115
estrogen analogs   66, 68, 72, 77–79
estrogen receptor   1–2, 7, 10, 13–14, 18, 38
estrogen replacement therapy 67–68, 70, 78

gene targeting   54, 61
glucocorticoid receptor   52, 55, 59
gluconeogenesis   56, 59

hypercalcemia   163, 175–176
hyperprolactinemic hypogonadism 125, 127, 129
hypogonadism   122–126, 134

idiopathic hypogonadotropic hypogonadism   127–128
implantation   108, 112–116

keratinocytes   138, 141, 147

mammary gland   86, 89–91

osteoblasts   68, 70, 78, 166–167, 175
osteocalcin   144–145, 151–152
osteoclasts   68, 71, 73, 78
osteopenia   122, 125, 129–130, 134
osteoporosis   121–122, 127, 132, 135
ovulation   108–110, 112–115

phosphorylation   31, 37–38, 40–43
primary hypogonadism   123
progestins   86, 88–95, 98, 101–102
protein kinases   31, 37–38, 41

rapid actions   162

tamoxifen   2, 9, 11, 13–14, 16, 19, 21, 29

transcription activation   30, 32, 35,
    37
tyrosine aminotransferase gene
    58–59

uterus   85–87, 89, 92

vitamin D   137–141, 143, 147,
    150, 153
vitamin D analogs   161, 164, 167,
    177
vitamin D receptor   138, 143, 146
vitamin D response element   146,
    152

Ernst Schering Research Foundation Workshop

Editors: Günter Stock
Ursula-F. Habenicht

Vol. 1
Bioscience ⇌ Society
Workshop Report
Editors: D. J. Roy, B. E. Wynne, R. W. Old

Vol. 2
Round Table Discussion on Bioscience ⇌ Society
Editor: J. J. Cherfas

Vol. 3
Excitatory Amino Acids and Second Messenger Systems
Editors: V. I. Teichberg, L. Turski

Vol. 4
Spermatogenesis – Fertilization – Contraception
Editors: E. Nieschlag, U.-F. Habenicht

Vol. 5
Sex Steroids and the Cardiovascular System
Editors: P. Ramwell, G. Rubanyi, E. Schillinger

Vol. 6
Transgenic Animals as Model Systems for Human Diseases
Editors: E. F. Wagner, F. Theuring

Vol. 7
Basic Mechanisms Controlling Term and Preterm Birth
Editors: K. Chwalisz, R. E. Garfield

Vol. 8
Health Care 2010
Editors: C. Bezold, K. Knabner

Vol. 9
Sex Steroids and Bone
Editors: R. Ziegler, J. Pfeilschifter, M. Bräutigam

Vol. 10
Nongenotoxic Carcinogenesis
Editors: A. Cockburn, L. Smith

Vol. 11
Cell Culture in Pharmaceutical Research
Editors: N. E. Fusenig, H. Graf

Vol. 12
Interactions Between Adjuvants, Agrochemical and Target Organisms
Editors: P. J. Holloway, R. T. Rees, D. Stock

Vol. 13
Assessment of the Use of Single Cytochrome P450 Enzymes
in Drug Research
Editors: M. R. Waterman, M. Hildebrand

Vol. 14
Apoptosis in Hormone-Dependent Cancers
Editors: M. Tenniswood, H. Michna

Vol. 15
Computer Aided Drug Design in Industrial Research
Editors: E. C. Herrmann, R. Franke

Vol. 16
Organ-Selective Actions of Steroid Hormones
Editors: D. T. Baird, G. Schütz, R. Krattenmacher

Supplement 1
Molecular and Cellular Endocrinology of the Testis
Editors: G. Verhoeven, U.-F. Habenicht